McGRAW-HILL NETWORKING AND TELECOMMUNICATIONS

Build Your Own
Trulove — *Build Your Own Wireless LAN* (with Projects)

Crash Course
Louis	*Broadband Crash Course*
Vacca	*I-Mode Crash Course*
Louis	*M-Commerce Crash Course*
Shepard	*Telecom Convergence,* Second Edition
Shepard	*Telecom Crash Course*
Bedell	*Wireless Crash Course*
Kikta/Fisher/Courtney	*Wireless Internet Crash Course*

Demystified
Harte/Levine/Kikta	*3G Wireless Demystified*
LaRocca	*802.11 Demystified*
Muller	*Bluetooth Demystified*
Evans	*CEBus Demystified*
Bayer	*Computer Telephony Demystified*
Hershey	*Cryptography Demystified*
Taylor	*DVD Demystified*
Bates	*GPRS Demystified*
Symes	*MPEG-4 Demystified*
Camarillo	*SIP Demystified*
Shepard	*SONET/SDH Demystified*
Topic	*Streaming Media Demystified*
Symes	*Video Compression Demystified*
Shepard	*Videoconferencing Demystified*
Bhola	*Wireless LANs Demystified*

Developer Guides
Guthery	*Mobile Application Development with SMS*
Richard	*Service and Device Discovery: Protocols and Programming*
Vacca	*I-Mode Crash Course*

Professional Telecom
Smith/Collins	*3G Wireless Networks*
Bates	*Broadband Teleco~ H-- ---* n
Collins	*Carrier Gr*
Harte	*Delivering*
Held	*Deploying (*
Minoli/Johnson/Minoli	*Ethernet-B*
Benner	*Fibre Chan~*

Bates	*GPRS*
Sulkin	*Implementing the IP-PBX*
Lee	*Lee's Essentials of Wireless*
Bates	*Optical Switching and Networking Handbook*
Wetteroth	*OSI Reference Model for Telecommunications*
Russell	*Signaling System #7*, Fourth Edition
Ohrtman	*Softswith: Architecture for VoIP*
Minoli/Johnson/Minoli	*SONET-Based Metro Area Networks*
Nagar	*Telecom Service Rollouts*
Louis	*Telecommunications Internetworking*
Russell	*Telecommunications Protocols,* Second Edition
Minoli	*Voice over MPLS*
Karim/Sarraf	*W-CDMA and cdma2000 for 3G Mobile Networks*
Bates	*Wireless Broadband Handbook*
Faigen	*Wireless Data for the Enterprise*

Reference

Muller	*Desktop Encyclopedia of Telecommunications,* Third Edition
Botto	*Encyclopedia of Wireless Telecommunications*
Clayton	*McGraw-Hill Illustrated Telecom Dictionary*, Third Edition
Pecar	*Telecommunications Factbook*, Second Edition
Russell	*Telecommunications Pocket Reference*
Radcom	*Telecom Protocol Finder*
Kobb	*Wireless Spectrum Finder*
Smith	*Wireless Telecom FAQs*

Security

Hershey	*Cryptography Demystified*
Nichols	*Wireless Security*

Telecom Engineering

Smith/Gervelis	*Cellular System Design and Optimization*
Rohde/Whitaker	*Communications Receivers*, Third Edition
Sayre	*Complete Wireless Design*
OSA	*Fiber Optics Handbook*
Lee	*Mobile Cellular Telecommunications*, Second Edition
Bates	*Optimizing Voice in ATM / IP Mobile Networks*
Roddy	*Satellite Communications*, Third Edition
Simon	*Spread Spectrum Communications Handbook*
Snyder	*Wireless Telecommunications Networking with ANSI-41*, Second Edition

BICSI

Network Design Basics for Cabling Professionals
Networking Technologies for Cabling Professionals
Residential Network Cabling
Telecommunications Cabling Installation

WI-FI HANDBOOK

Wi-Fi Handbook

Building 802.11b
Wireless Networks

Frank Ohrtman
Konrad Roeder

McGraw-Hill
New York   Chicago   San Francisco   Lisbon
London   Madrid   Mexico City   Milan   New Delhi
San Juan   Seoul   Singapore   Sydney   Toronto

The **McGraw·Hill** Companies

Library of Congress Cataloging-in-Publication Data

Ohrtman, Frank.
   Wi-Fi handbook: building 802.11b wireless networks / Frank Ohrtman,
Konrad Roeder.
      p.   cm.
   Includes Index
   ISBN 0-07-141251-4 (alk. paper)
   1. IEEE 802.11 (Standard)   2. Internet service providers—Standards.
   3. Telecommuncation—Technological innovations—United States.
   I. Roeder, Konrad.   II. Title

   TK5105.5668.036   2003
   004.67'8—dc21   2003044574

1 2 3 4 5 6 7 8 9 0   DOC/DOC   0 9 8 7 6 5 4 3

ISBN 0-07-141251-4

*The sponsoring editor for this book was Stephen S. Chapman and the production
supervisor was Sherri Souffrance. It was set in Century Schoolbook by MacAllister
Publishing Services, LLC.*

Printed and bound by RR Donnelley.

McGraw-Hill books are available at special quantity discounts to use as premiums and
sales promotions, or for use in corporate training programs. For more information,
please write to the Director of Special Sales, Professional Publishing, McGraw-Hill,
Two Penn Plaza, New York, NY 10121-2298. Or contact your local bookstore.

This book is dedicated to all the WISP and other wireless pioneers that are blazing new trails in wireless telecommunications.

CONTENTS

Contents

Contents

PREFACE

The Telecommunications Act of 1996 has failed to bring competition to the local loop of the North American telecommunications market. Despite its provisions ordering the opening of telecommunications facilities (switching and access) to competitors, less than 10 percent of Americans have any choice in their local telephone service provider. Despite its widespread popularity in the residential market, Internet access remains for most Americans mired in dialup access only. Fewer than 10 percent of American households have broadband Internet access. Many service providers consider their *Digital Subscriber Line* (DSL) or cable TV Internet access to be broadband where the access speeds might top out at 256 Kbps. If a service provider does not deem it economically viable to deploy the necessary infrastructure for a given neighborhood, then those subscribers are barred from the economic benefits of broadband Internet access.

The emergence of 802.11 technologies holds the promise of both introducing competition to the local loop telephone service while delivering true broadband Internet access. Economists Robert Crandall and Charles Jackson estimate ubiquitous broadband Internet access could bring a $500 billion annual economic benefit to the American economy. The chief barrier to achieving that benefit is the "last-mile bottleneck" of the current telecommunications infrastructure, consisting of either copper pair telephone lines or coaxial cable from the cable TV companies. These technologies cannot come close to offering the 11 Mbps bandwidth prescribed in the 802.11 specification. Furthermore, because 802.11 is wireless Ethernet, its cost of deployment is considerably less than that of telephone wires or coaxial cable. The ease and relative economy of deploying 802.11 should dictate a relatively rapid acceptance by entrepreneurs in rolling out telecommunications services.

However, 802.11 has sparked a number of objections. This book was written to explain how these objections are not the brick walls to adoption they may appear to be. Those technical objections include security, *quality of service* (QoS), and range. Other objections revolve around the economic and regulatory issues associated with 802.11

technologies, one of which concerns unlicensed spectrum. People want to know how long the spectrum will be unlicensed or if it will succumb to a "tragedy of the commons," in that so many users occupy the spectrum that it becomes unusable. Recent announcements by the *Federal Communications Commission* (FCC) point to an expansion of unlicensed spectrum. New legislation being introduced in Congress indicates support for 802.11 technologies and services in the legislature.

As objections to 802.11 technologies are worked through, a great wave of new innovation will come crashing into the telecommunications markets, creating many unforeseen economic opportunities. It is the hope of the authors that this book can be one small signpost on the path to prosperity for wireless entrepreneurs.

ACKNOWLEDGMENTS

This book would not have been possible without the generous donation of time and talent from Mike Schmidt, Jackie Peterson, Richard Siber, Geri Mitchell, Karim Damji, Chris Fine, Patrick Sorqvist, Priyank Garg, Rushabh Doshi, Majid Malek, Russell Greene, Maggie Cheng, Rob Flickinger, Mathew Gast, Brent Bierstedt, Fred Goldstein, Plamen Nedeltchev, Jim Zyhren, Al Petrick, Tim Pozar, Friederick Kaemmerer, James and Ruth LaRocca, Michael Collins, Doug Haslam, Kathryn Korostoff, Charles Jackson, Robert Crandall, and Dale Hatfield.

As with any great work of literature, this book could not have been written without the cheerful assistance of the barristas Jennifer Le Grand, Heather Thompson, and Robin Balchen at St. Mark's coffee house and Maria Steed at Diedrich's coffee house, both in Denver, Colorado.

Introduction

In the late 1990s, the telecommunications boom went bust because new market entrants, known as *Competitive Local Exchange Carriers* (CLECs), were forced to compete with *Incumbent Local Exchange Carriers* (ILECs) on the same financial terms of the incumbents. The failure of the CLECs resulted in a net investment loss of approximately $3 trillion, adversely affecting capital markets and severely depressing the overall telecommunications economy as well as saddling subscribers with artificially high rates.

The Telecommunications Act of 1996 aimed to introduce competition in the local loop by legally requiring incumbents to lease space on their switches and provide access to their subscribers to any and all competitors. New market entrants found themselves stonewalled in the courts by the incumbents when attempting to gain legal access to the incumbents' facilities. Once legal access was gained to the incumbents' switching facilities, the incumbents conveniently forgot the orders or otherwise sabotaged the operations of the CLECs in the incumbents' switching facilities.

Given the astronomical expense of deploying a conventional, but alternative network or the legal obstacle of gaining access to the *Public Switched Telephone Network* (PSTN), it is not surprising that seven years after the passage of the Telecommunications Act of 1996, only 9 percent of American residential phone lines are handled by competitive carriers. Given this dismal figure, it is clear that regulatory agencies such as the *Federal Communications Commission* (FCC) and the utilities commissions of the 50 states have failed to adequately enforce either the letter or spirit of the Telecommunications Act in regards to introducing competition in the local loop.

A competitive local loop environment has two apparently insurmountable obstacles: (1) the high cost of Class 4 and Class 5 switches and (2) gaining access to the local loop network. At the time of this writing, despite the guarantees contained in the Telecommunications Act of 1996, it appears obvious that competition will never come *in* the local loop but will have to come *to* the local loop in the form of an alternative network. The only way consumers will enjoy the benefits of competition in the local loop is when alternative technology in switching and access offers a competitor less barriers to entering and exiting the telecommunications market. If telecommunications consumers are supposed to enjoy the benefits of competition in their local loop, the ability to bypass the switching architecture to gain access (via copper wires from the telephone company) must be offered.

Telecommunications Networks— The Need for an Alternative Form of Access

An understanding of the PSTN is best grasped by examining its three major components: access, switching, and transport (see Figure 1-1). Each element has evolved over the 100-plus-year history of the PSTN. *Access* pertains to how a user accesses the network. *Switching* refers to how a call is switched or routed through the network. *Transport* describes how a call travels or is transported over the network. This network was designed to handle voice. Eventually, data was introduced onto this network. As data traffic on the PSTN grew, high-capacity users found it inadequate. These subscribers then moved their data traffic to data-specific networks. Many data users find themselves limited to an infrastructure that is dependent on wires, whether they are using fiber-optic cable, coaxial cable, or twisted-pair copper wire. Although wireless communication is not new (forms of radio communication have been in use for almost a century), using wireless communication to bypass wired monopolies is now a practical opportunity for subscribers of both voice and data

Figure 1-1
The three components of a telephone network: access, switching, and transport

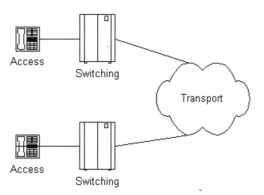

services. The primary form of bypass is the use of cellular phones. The wireless technology 802.11b also holds great promise in delivering broadband data (up to 11 Mbps).

Access

Access refers to how a user accesses the telephone network. For most users, access is gained to the network via a telephone handset. Transmission and reception occurs via diaphragms where the mouthpiece converts the air pressure of voice into an analog electromagnetic wave for transmission to the switch. The earpiece performs this process in reverse. The most sophisticated aspect of the handset is its *Dual-Tone Multifrequency* (DTMF) function, which signals the switch by tones. The handset is usually connected to the central office (where the switch is located) via copper wire known as *twisted pair*, because, in most cases, it consists of a twisted pair of copper wire. The stretch of copper wire connects the telephone handset to the central office. Everything that runs between the subscriber and the central office is known as the *outside plant*. Telephone equipment at the subscriber end is called *customer premise equipment* (CPE). One of the chief reasons the majority of subscribers have no choice in local service providers is the prohibitive expense of deploying any alternative to the copper wire that now connects them to the network. Secondly, gaining right-of-way across properties to reach subscribers borders on the impossible both in legal and economic terms.

Switching

The PSTN is a star network; that is, every subscriber is connected to another via at least one, if not many, hubs known as *offices*. Those offices contain switches. Very simply, local offices are used for local service connections and tandem offices are used for long-distance service. Local offices, better known as *central offices*, use Class 5 switches and tandem offices use Class 4 switches. Figure 1-2 details the relationship between Class 4 and Class 5 switches. A large city might have several central offices. Denver (population 2 million), for example, has approximately 40 central offices. Central offices in a

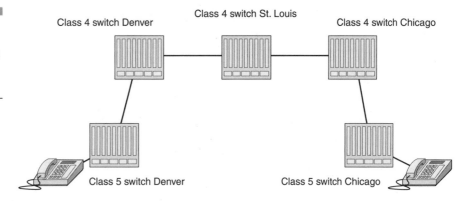

Figure 1-2
The relationship between Class 4 and Class 5 switches.

large city often take up much of a city block and are recognizable as large brick buildings with no windows.

Transport

The PSTN was built at great expense over the course of more than a century. Developers have been obsessed over the years with getting the maximum number of conversations transported at the least cost in infrastructure possible. Imagine an early telephone circuit running from New York to Los Angeles. The copper wire, repeaters, and other mechanisms involved in transporting a conversation this distance were immense for the time. Hence, the early telephone engineers and scientists had to find ways to get the maximum number of conversations transported over this network. Through much research, different means were developed to achieve the maximum efficiency from the copper wire infrastructure. Many of those discoveries translated on technologies that worked equally well when fiber-optic cable came into the market. The primary form of transport in the PSTN has been *time-division multiplexing* (TDM). In the 1990s, long-distance service providers (*interexchange carriers* [IXCs]) and local service providers (*Local Exchange Carriers* [LECs]) migrated those transport networks to *Asynchronous Transfer Mode* (ATM). ATM is a means of transport from switch to switch. The emergence of *Internet Protocol* (IP) backbones is drawing much of the traffic off ATM networks and moving it to IP networks.

Replacing the PSTN One Component at a Time

The three components of the PSTN are being replaced in the free market via substitution by other technologies and changes in the regulatory atmosphere. The *Memorandum of Final Judgement* (MFJ) of 1984 opened the transport aspect of the PSTN to competition. This gave rise to an explosion in the number of long-distance service providers in the United States. The bandwidth glut of 2000 has driven down the cost of long-distance transport.

The Telecommunications Act of 1996 was intended to further the reforms brought on by the MFJ of 1984, but it has failed to do so. The act specified how incumbent telephone companies were to open their switches to competitors. The incumbents stalled this access first by legal maneuvering and then by outright sabotage. The same tactics were employed in blocking competitive access to the access side of their networks. A technology known as *softswitch* offers a technology bypass of the PSTN switches. This still leaves the *last mile* (also known as the *first mile*) under the control of the incumbent service providers.

A new technology known as 802.11b and its associated variants offer the possibility for service providers to bypass the incumbents' local loop to deliver service to the last mile. The applications for 802.11b began with enterprise and government networks and have migrated to home networks. In the year prior to this writing, the industry experienced an explosion in the sales of wireless network products. As this technology becomes more popular, subscribers will gain confidence in wireless technologies and their related services, and will increasingly "cut the wire" to incumbent wired service providers.

This book provides a roadmap for cutting those wires. It provides an introduction to 802.11b and its related protocols (802.16, 802.11a, 802.11g, 802.11i, and so on). It offers extensive evidence of the myriad applications of 802.11b currently in operation with guidelines for implementing 802.11b. A series of case studies offer evidence of the viability of 802.11b and related wireless technologies.

802.11 Works—An Overview of the Installation and Operation of Wireless Networks

The evidence of successful deployments of wireless networks for both data and voice applications raises questions as to whether this technology could be deployed as an alternative to the PSTN. If it carries data and voice competently, why should a business or residence continue to subscribe to expensive (and often monopolistic) wireline services? The emergence of *voice over Internet Protocol* (VoIP) and its associated infrastructure technologies (for example, softswitch) reduces the transmission of voice to the simple routing and transportation of data packets. Hence, it is no longer necessary for a subscriber to contract with a telephone company for local or long-distance voice services.

Despite the popularity of the Internet and its myriad services, incumbent service providers, telephone companies, and cable TV companies have failed to offer a ubiquitous broadband Internet service. If broadband Internet access were as ubiquitous as telephone service was at the time of this writing, the American economy, for example, would reap a $500 billion annual benefit.[1] This book explores wireless architectures that will potentially compete with the PSTN for the delivery of voice and data services.

Objections to Wireless Networks

The position that wireless technologies will replace the PSTN is met with a number of objections. Primarily, these objections are focused on *quality of service* (QoS) issues, the security of the wireless

---

[1]Robert Crandall and Charles Jackson, "The $500 Billion Opportunity: The Potential Economic Benefit of Widespread Diffusion of Broadband Internet Access," *Criterion Economics* (July 2001): 69.

network, limitations in the range of the delivery of the service, and the availability of bandwidth. This book overcomes these objections.

Quality of Service (QoS)

One of the primary concerns about wireless data delivery is that, like the Internet over wired services, QoS is inadequate. Contention with other wireless services, lost packets, and atmospheric interference are recurring problems for 802.11b and its associated wireless protocols as alternatives to the PSTN. QoS is also related to the ability of a *wireless Internet service provider* (WISP) to accommodate voice on its network. The PSTN cannot be replaced until an alternative, competent replacement for voice over copper wire is available (see Figure 1-3).

Figure 1-3
An overview of a broadband wireless alternative to the PSTN

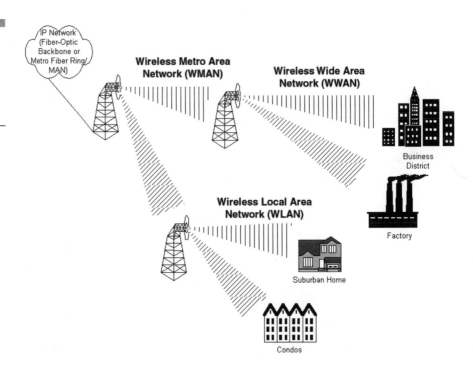

Security

The press has been quick to report on weaknesses found in wireless networks. 802.11b has two built-in basic network security mechanisms: the *service set identifier* (SSID) and *Wireless Equivalency Privacy* (WEP). These measures may be adequate for residences and small businesses, but they are inadequate for entities that require stronger security. A number of measures that will provide the necessary level of security for the subscriber can be added to those wireless networks. This book provides suggestions for deploying industrial-grade security.

Range

In most applications, 802.11b offers a range of about 100 meters. So how, you might ask, will that technology offer the range to compete with the PSTN? Range is a function of antenna design and power, but mostly antenna design. With the right antenna, the range of 802.11 is extended to tens of miles.

The Economic Advantage of 802.11

Every *information technology* (IT) manager and manager of any alternative service provider must carefully weigh both the *return on investment* (ROI) and the *net present value* (NPV) of a new technology when deciding on investing in new platforms. Is a wireless network less expensive to purchase and operate than a wired network? What about the convergence of voice and data on one network? What about apparent intangibles such as worker productivity on wired versus wireless networks? This book offers practical examples of ROI and NPV problems to help solve these dilemmas.

For service providers, wireless technologies pose a potential cost-effective solution in that they do not require right-of-way across private or public property to deliver service to the customer. Many businesses cannot receive broadband data services as no fiber-optic cable runs to their building(s). The cost of securing permission to dig a trench through another property and running the requisite cable is prohibitive. With 802.11b and its associated technologies, it is possible to merely beam the data flow to that building. This solution carries over to the *small office/home office* (SOHO) market in that the data flow can be beamed to homes and small businesses in places where no fiber-optic or other high-bandwidth service exists.

The Regulatory Aspects of Wireless Networks

What are the regulatory concerns for a WISP when deploying a wireless enterprise network? The FCC addresses wireless services in what is popularly known as *Part 15*. Wireless data requires spectrum on which to transmit over the airwaves at a given frequency. 802.11 and most of its associated protocols operate on what is known as *unlicensed spectrum*. Unlicensed spectrum does not require the operator to obtain an exclusive license to transmit on a given frequency in a given region. Unlike the operators of radio stations or cellular telephone companies, a WISP, public or private, is transmitting for free. Assuming WISPs ultimately compete with cell phone companies for subscribers, a WISP that utilizes 802.11 technologies may find itself at a strong advantage over *third-generation* (3G) networks. 802.11 delivers wireless data up to 11 Mbps on cost-free unlicensed spectrum, whereas 3G delivers bandwidth at approximately 128 Kbps over very expensive licensed spectrum. Will the unlicensed spectrum remain free of charge? How should conflicts on the airwaves be settled? How is the public best served in the commons of the frequency spectrum?

Improved Quality of Life with Wireless Networks

When deployed as a broadband IP network solution, 802.11b will enable an improved standard of living in the form of telecommuting, lower real-estate prices, and improved quality of life. A wave of opportunity for wireless applications is in the making. Most of it lies in the form of broadband deployment. The potential for better living through telecommunications lies largely with the ubiquitous availability of broadband.

Only approximately 9 percent of American households have access to broadband Internet. Incumbent service providers have failed to expand that figure given the cost of their wired infrastructure and right-of-way legal barriers. Given the relative low cost of delivering wireless data to a business or residence, 802.11b technologies offer a very convenient alternative to conventional technologies in deploying broadband Internet to businesses and residences around the world.

Disruptive Technology

In the business book *The Innovator's Dilemma* (2000), Clayton Christensen describes how disruptive technologies have precipitated the failure of leading products as well as their associated and well-managed firms. Christensen defines criteria to identify disruptive technologies regardless of their market. Such technologies have the potential to replace mainstream technologies as well as their associated products and principal vendors. Disruptive technologies, abstractly defined by Christensen, are "typically cheaper, simpler, smaller, and, frequently, more convenient" than their mainstream counterparts.

Wireless technology, compared to incumbent wired networks, is a disruptive technology. For the competitive service provider, 802.11b is "cheaper, simpler, smaller, and, frequently, more convenient" than copper wire and its associated infrastructure. In order for a technology to be truly disruptive, it must disrupt an incumbent vendor or service provider. Some entity must go out of business before a technology can be considered disruptive. Although it is too early to point out the incumbent service providers driven out of business by 802.11b, its technologies could be potentially disruptive to incumbent telephone companies. The migration of wireline telephone traffic from ILEC to cellular is a powerful example of this trend.

Conclusion

This book describes how wireless technologies meet or exceed the performance parameters of wired last-mile networks and pose a potentially disruptive scenario for telephone service providers. In a market economy, it is inevitable that if competition cannot come in the local loop, it will surely come to the local loop. Given that wireless technologies could potentially match the last mile in terms of QoS, security, range, bandwidth, and economics, 802.11b provides the crucial avenue for competitive service providers to enter telecommunications markets worldwide. Delivering broadband Internet to homes and small businesses has many societal benefits. As a result of the sloth of incumbent service providers in deploying broadband Internet access to the last mile, wireless applications present what is probably the fastest avenue in delivering a huge economic surplus to society.

How Does 802.11 Work?

What, technically speaking, is 802.11b and how does it relate to IEEE 802.11? This chapter covers the technology of transmitting data over the airwaves, the process of that transmission, and the topologies and components of wireless networks. Thousands of enterprises worldwide are cutting the wires to their *local area networks* (LANs) to enjoy greater productivity from their unwired work force. 802.11b also has the potential to save money on infrastructure (wiring buildings for networks) and telecommunications services.

How Does Wireless Fidelity (Wi-Fi) Work?

A networked desktop computer is connected to a larger network (a LAN, *wide area network* [WAN], or the Internet) via a network cable to a hub, router, or switch. The computer's *network interface card* (NIC) sends zeros and ones down the cable by changing the voltage on the wires from +5 volts to -5 volts in a prearranged cadence. Wi-Fi simply replaces these cables with small, low-powered two-way radios. Instead of changing the voltage on a wire, it encodes the zeros and ones by laying an alternating radio signal over a constant existing signal in a prearranged cadence. The alternating signal encodes zeros and ones on the radio waves. The 802.11b specification allows for the wireless transmission of approximately 11 Mbps of raw data at distances up to a few hundred feet over the 2.4 GHz unlicensed band. The distance depends on impediments, materials, and line of sight.

Many people might ask, "So what?" It means a PC user can install a $70 PC card in his or her laptop or *personal digital assistant* (PDA) and be connected to the Internet or corporate network as if he or she was still tied to a desk. Enterprises have been quick to adopt this technology based on the following factors:

- Wiring a building for voice and data is expensive.

- It improves worker productivity by allowing mobility within a building or corporate campus.

- It does not require right-of-way agreements to bring service to a business.

- It is independent of distance limitation to the central office.
- It is relatively free of federal, state, and local regulations.

A *wireless local area network* (WLAN) installation usually uses one or more *access points* (APs), which are dedicated stand-alone hardware with typically more powerful antennas. Figure 2-1 illustrates a WLAN. In addition to servicing enterprise networks, 802.11b has become the most popular standard for public short-range networks, known as *hot spots*, found at airports, hotels, conference centers, and coffee shops and restaurants. Several companies currently offer paid hourly, session-based, or unlimited monthly access via their deployed networks around the United States and internationally.[1]

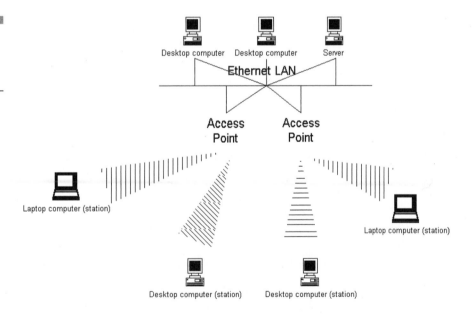

Figure 2-1
A WLAN on an enterprise network

[1]"OK, What Is Wi-Fi?" a white paper from Wi-Fi Consulting, www.wificonsulting.com/Wi-Fi101/wifi101.htm.

How Is Data Transmitted via Wireless Technology?

The 802.11 standard provides for two *radio frequency* (RF) variations (as opposed to infrared) of the *physical* (PHY) layer. These include *direct sequence spread spectrum* (DSSS) and *frequency-hopping spread spectrum* (FHSS). Both of these were designed to comply with the *Federal Communications Commission* (FCC) regulations (FCC 15.247) for operation in the 2.4 GHz band, which is an unlicensed spectrum. 802.11b uses DSSS.

DSSS systems use technology similar to *Global Positioning System* (GPS) satellites and some types of cell phones. Each information bit is combined with a longer *pseudorandom numerical* (PN) in the transmission process. The result is a high-speed digital stream, which is then modulated onto a carrier frequency using *differential phase-shift keying* (DPSK). Figure 2-2 illustrates how data is modulated with a PN sequence for wireless transmission.[2]

As illustrated in Figure 2-2, DSSS works by taking a data stream of zeros and ones and modulating it with a second pattern—the chipping sequence. The sequence is also known as the *Barker code*, which is an 11-bit sequence (10110111000). The chipping or spreading code is used to generate a redundant bit pattern to be transmitted, and

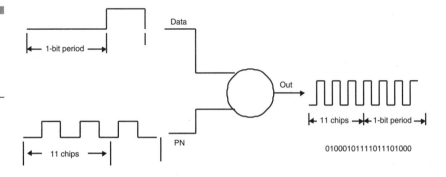

Figure 2-2
Digital modulation of data with PN sequence

1-bit period

11 chips

Data

PN

Out

11 chips — 1-bit period

01000101111011101000

---

[2]Jim Zyren and Al Petrick, "IEEE 802.11 Tutorial," a white paper from Wireless Ethernet, www.wirelessethernet.org/downloads/IEEE_80211_Primer.pdf.

the resulting signal appears as wideband noise to the unintended receiver. One of the advantages of using spreading codes is that even if one or more of the bits in the chip are lost during transmission, statistical techniques embedded in the radio can recover the original data without the need for retransmission. The ratio between the data and width of spreading code is called *processing gain*. It is 16 times the width of the spreading code and increases the number of possible patterns to 64,000 (2^{16}), reducing the chances of cracking the transmission.

The DSSS signaling technique divides the 2.4 GHz band into fourteen 22 MHz channels, of which 11 adjacent channels overlap partially and the remaining 3 do not overlap. Data is sent across one of these 22 MHz channels without hopping to other channels, causing noise on the given channel. To reduce the number of retransmissions and noise, chipping is used to convert each bit of user data into a series of redundant bit patterns called *chips*. The inherent redundancy of each chip, combined with spreading the signal across the 22 MHz channel, provides the error checking and correction functionality to recover the data. Spread spectrum products are often interoperable because many are based on the IEEE 802.11 standard for wireless networks. DSSS is used primarily in interbuilding LANs, as its properties are fast and far reaching.[3]

At the receiver, a matched filter correlator is used to remove the PN sequence and recover the original data stream. At a data rate of 11 Mbps, DSSS recéivers use different PN codes and a bank of correlators to recover the transmitted data stream. The high-rate modulation method is called *complimentary code keying* (CCK).

As illustrated in Figure 2-2 the PN sequence spreads the transmitted bandwidth of the resulting signal (hence the term *spread spectrum*) and reduces peak power. The total power remains unchanged. Upon receipt, the signal is correlated with the same PN sequence to reject narrowband interference and recover the original binary data. Regardless of whether the data rate is 1, 2, 5.5, or 11 Mbps, the channel bandwidth is about 20 MHz for DSSS systems.

---

[3]Plamen Nedeltchev, "WLANs and the 802.11 Standard," a white paper from Cisco Systems, www.cisco.com/warp/public/784/packet/jul01/pdfs/whitepaper.pdf

Figure 2-3
The spreading
of spectrum for
transmission in
DSSS or FHSS

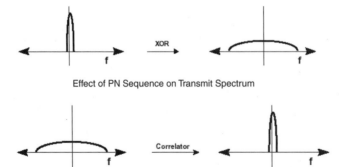

Effect of PN Sequence on Transmit Spectrum

Received Signal is Correlated with PN to Recover Data and Reject Interference

The Significance of Spread Spectrum Radio

One of the basic technologies underlying the IEEE 802.11 series of standards is spread spectrum radio. The fundamental concept of spread spectrum radio is the use of a wider frequency bandwidth than that needed by the information that is transmitted. Using extra bandwidth would seem to be wasteful, but it actually results in several benefits, including reduced vulnerability to jamming, less susceptibility to interference, and coexistence with narrowband transmissions. Several spread spectrum techniques are available, including time hopping, frequency modulation, FHSS, DSSS, and hybrids of these.

FHSS and DSSS are not modulation techniques, but methods of distributing a radio signal across bandwidth. In addition to spreading the signal across a frequency band, spread spectrum systems modulate the signal. Modulation is the variation of a radio signal to convey information. The base signal is called the *carrier*. The variation may be based on the strength (*amplitude modulation* [AM]), frequency, or phase (frequency offset) of the signal. The modulation technique directly affects the data rate. Higher data rate modulations are generally more complex and expensive to implement. Modulations resulting in higher data rates pack more information in the same bandwidth. Small disruptions in the signal cause the degradation of more data. This means that the signal must have a higher *signal-to-noise ratio* (SNR) at the receiver to be effectively processed. Because a radio signal is stronger the closer it is to the source, the

SNR decreases with distance. This is why higher-speed systems have less range. Examples of modulation techniques used in the IEEE 802.11 series of specifications include *binary phase-shift keying* (BPSK), *quadrature phase-shift keying* (QPSK), *Gaussian frequency-shift keying* (GFSK), and CCK.

802.11 Variants

In 1997, the IEEE adopted IEEE Standard 802.11-1997, the WLAN standard. This standard defines the *Medium Access Control* (MAC) and PHY layers for a LAN with wireless connectivity. It addresses local area networking where the connected devices communicate over the air to other devices that are within close proximity to each other. Figure 2-1 illustrates this.

The industry group *Wireless Ethernet Compatibility Alliance* (WECA) certifies its members' equipment as conforming to the 802.11b standard and enables compliant hardware to be certified as Wi-Fi compatible. This is an attempt to guarantee intercompatibility between hundreds of vendors and thousands of devices. Table 2-1 lists the variants of 802.11 and provides an overview of the relationship between 802.11b with other 802.11 variants.

Table 2-1

IEEE 802.11
Variants

802.11 Variant	Description
802.11a	Created a standard for WLAN operations in the 5 GHz band with data rates of up to 54 Mbps. Published in 1999.
802.11b	Created a standard (also known as Wi-Fi) for WLAN operations in the 2.4 GHz band with data rates of up to 11 Mbps. Published in 1999. Products based on 802.11b include public-space Internet kiosks, WLAN services such as Wayport, and wireless home networking products such as the Macintosh AirPort.
802.11c	Provided documentation of 802.11-specific MAC procedures to the *International Organization for Standardization/International Electrotechnical Commission* (ISO/IEC) 10038 (IEEE 802.1d) standard. Work has completed.

Table 2-1 cont.

IEEE 802.11
Variants

802.11 Variant	Description
802.11d	Publishing definitions and requirements to enable the 802.11 standard to operate in countries that are not currently served by the standard.
802.11e	Attempting to enhance the 802.11 MAC to increase the *quality of service* (QoS) possible. Improvements in capabilities and efficiency are planned to allow applications such as voice, video, or audio transport over 802.11 wireless networks.
802.11f	Developing recommended practices for implementing the 802.11 concepts of APs and *distributed systems* (DSs). The purpose is to increase compatibility between AP devices from different vendors.
802.11g	Developing a higher-speed PHY extension to the 802.11b standard while maintaining backward compatibility with current 802.11b devices. The target data rate for the project is at least 20 Mbps.
802.11h	Enhancing the 802.11 MAC and 802.11a PHY to provide network management and control extensions for spectrum and transmit power management in the 5 GHz band. This will allow regulatory acceptance of the standard in some European countries.
802.11i	Enhancing the security and authentication mechanisms of the 802.11 standard.
802.1x	Also aimed at enhancing security of 802.11b.

Source: Wave Report, November 29, 2001.

FHSS (802.11a)

Spread spectrum radio techniques originated in the U.S. military in the 1940s. The unlikely co-patent holders on spread spectrum technology are the actress Hedy Lamar and musician George Antheil. Lamar had been married to a German arms dealer and fled Germany as the Nazis came to power. One of Antheil's techniques involved the use of player pianos. These two came together to create one of the twentieth century's most influential radio technologies.

The military had started to use radio as a remote control mechanism for torpedoes, but this technique suffered from a vulnerability to jamming. Aware of this, Lamar suggested to Antheil that the radio signal should be distributed randomly over time across a series of frequencies. The transmission on each frequency would be brief and make the aggregate less susceptible to interruption or jamming. The problem was synchronizing the transmitter and receiver to the frequency being used at any point in time. Antheil used his musical expertise to design a synchronization mechanism using perforated paper rolls like those found in player pianos.

Lamar and Antheil were awarded patent number 2,292,387 and gave the rights to the U.S. Navy in support of the war effort. Although the Navy did not deploy the technology, engineers at Sylvania Electronic Systems applied electronic synchronization techniques to the concept in the late 1950s. The U.S. military began using these systems for secure communications in the early 1960s. The spread spectrum technique spawned from the work of Hedy Lamar and George Antheil is what we now call FHSS.

Local authorities also regulate the hopping rate. In North America, the hopping rate is set at 2.5 hops per second with each transmission occupying a channel for less than 400 milliseconds. Channel occupancy is also called *dwell time*. In 2001, the FCC proposed to amend its Part 15 rules to allow adaptive hopping techniques to be used. This rulemaking is designed to reduce interference with other systems operating the 2.4 GHz frequencies. Studies have shown that up to 13 IEEE 802.11 FHSS systems can be co-located before frequency channel collisions become an issue.[4]

DSSS

DHSS systems mix high-speed bit patterns with the information being sent to spread the RF carrier. Each bit of information has a redundant bit pattern associated with it, effectively spreading the

---

[4]James and Ruth LaRocca, *802.11 Demystified* (New York: McGraw-Hill, 2002), 124–126.

signal over a wider bandwidth. These bit patterns vary in length and in the rate at which they are mixed into the RF carrier. They are called *chips* or *chipping codes* and range from 11 bits to extremely long sequences. The speed at which they are transmitted is called the *chipping rate*. To an observer, these sequences appear to be noise and are also called *pseudorandom noise codes* (Pncodes). Pncodes are usually introduced into the signal through the use of hardware-based shift registers, and the techniques used to introduce them are divided into several groups including Barker codes, Gold codes, M-sequences, and Kasami codes.

These spreading codes also allow the use of statistical recovery methods to repair damaged transmissions. Another side effect of spreading the signal is lower spectral density—that is, the same amount of signal power is distributed over more bandwidth. The effect of a less spectrally dense signal is that it is less likely to interfere with spectrally dense narrowband signals. Narrowband signals are also less likely to interfere with a DSSS signal because the narrowband signal is spread as part of the correlation function at the receiver.

The frequency channel in IEEE 802.11 DSSS is 22 MHz wide. This means that it supports three nonoverlapping channels for operation. This is why only three IEEE 802.11b DSSS systems can be colocated.

In addition to spreading the signal, modulation techniques are used to encode the data signal through predictable variations of the radio signal. IEEE 802.11 specifies two types of DPSK modulation for DSSS systems. The first is BPSK and the second is QPSK. *Phase-shift keying* (PSK), as the name implies, detects the phase of the radio signal. BPSK detects a 180-degree inversion of the signal, representing a binary 0 or 1. This method has an effective data rate of 1 Mbps. QPSK detects 90-degree phase shifts. This doubles the data rate to 2 Mbps. IEEE 802.11b adds CCK and *packet binary convolutional coding* (PBCC), which provide data rates up to 11 Mbps.[5]

---

[5]Ibid., 126–128.

Orthogonal Frequency Division Multiplexing (OFDM) and IEEE 802.11a

IEEE 802.11a (5 GHz) uses OFDM as its frequency management technique and adds several versions of *quadrature amplitude modulation* (QAM) in support of data rates up to 54 Mbps. In 1970, Bell Labs patented OFDM, which is based on a mathematical process called *Fast Fourier Transform* (FFT). FFT enables 52 channels to overlap without losing their individuality or orthoganality. Overlapping channels is a more efficient use of the spectrum and enables them to be processed at the receiver more efficiently. IEEE 802.11a OFDM divides the carrier frequency into 52 low-speed subcarriers. Forty-eight of these carriers are used for data and four are used as pilot carriers. The pilot subcarriers allow frequency alignment at the receiver.

One of the biggest advantages of OFDM is its resistance to multipath interference and delay spread. Multipath is caused when radio waves reflect and pass through objects in the environment. Radio waves are attenuated or weakened in a wide range depending on the object's materials. Some materials (such as metal) are opaque to radio transmissions. As you can see, a cluttered environment would be very different from an open warehouse environment for radio wave transmission and reception. This environmental variability is why it is so hard to estimate the range and data rate of an IEEE 802.11 system. Because of reflections and attenuation, a single transmission can be at different signal strengths and from different directions depending on the types of materials it encounters. This is multipath. IEEE 802.11a supports data rates from 6 to 54 Mbps. It utilizes BPSK, QPSK, and QAM to achieve the various data rates.

Delay spread is associated with multipath. Because the signal is traveling over different paths to the receiver, the signal arrives at different times. This is delay spread. As the transmission rate increases, the likelihood of interference from previously transmitted signals increases. Multipath and delay spread are not much of an issue at data rates less than 3 or 4 Mbps, but some sort of mechanism is required as rates increase to mitigate the effect of multipath and delay spread. In IEEE 802.11b, it is CCK modulation. In 802.11a, it is OFDM. The IEEE 802.11g specification also uses OFDM as its frequency management mechanism.[6]

The adoption and refinement of advanced semiconductor materials and radio transmission technologies for IEEE 802.11 provides a solid basis for the implementation of higher-level functions. The next step up the protocol ladder is the definition of access functionality. Without structured access, the physical medium would be unusable.[7]

Orthogonal Frequency Division Multiplexing

802.11a is based on OFDM. OFDM is not a new technique. Most of the fundamental work was done in the late 1960s, and U.S. patent number 3,488,445 was issued in January 1970. Recent *Digital Subscriber Line* (DSL) work (such as *High-bit-rate Digital DSL* [HDSL], *Very high-bit-rate DSL* [VDSL], and *Asymmetrical DSL* [ADSL]) and wireless data applications have rekindled interest in OFDM, especially now that better signal-processing techniques make it more practical. OFDM does, however, differ from other emerging encoding techniques such as *Code Division Multiple Access* (CDMA) in its approach. CDMA uses complex mathematical transforms to put multiple transmissions onto a single carrier; OFDM encodes a single transmission into multiple subcarriers. The mathematics underlying the code | division in CDMA is far more complicated than in OFDM. OFDM devices use one wide frequency channel by breaking it up into several component subchannels. Each subchannel is used to transmit data. All the low subchannels are then multiplexed into one code | division combined channel.

---

[6]Ibid., 131.

[7]Ibid., 99, 129–131.

Carrier Multiplexing

When network managers solicit user input on network build-outs, one of the most common demands is for more speed. The hunger for increased data transmissions has driven a host of technologies to increase speed. OFDM takes a qualitatively similar approach to *Multilink Point-to-Point Protocol* (MPPP)—when one link isn't enough, use several in parallel.

OFDM is closely related to plain old *frequency division multiplexing* (FDM). Both divide the available bandwidth into slices called carriers or subcarriers and make those carriers available as distinct channels for data transmission. OFDM boosts throughput by using several subcarriers in parallel and multiplexing data over the set of subcarriers.

Traditional FDM was widely used by first-generation mobile telephones as a method for radio channel allocation. Each user was given an exclusive channel, and guard bands were used to ensure that spectral leakage from one user did not cause problems for users of adjacent channels.[8]

Power Management and Time Synchronization

In addition to the *carrier sense multiple access with collision avoidance* (CSMA/CA) control frames (*Request to Send* [RTS], *Clear to Send* [CTS], *acknowledgment* [ACK], and contention polling), the MAC also provides control frames for power management and time synchronization. APs provide a time synchronization beacon to associated stations in an infrastructure *basic service set* (BSS). In an *independent BSS* (IBSS) where stations are operating as peers, an algorithm is defined that enables each station to reset its time when

---

[8]Matthew Gast, *802.11 Wireless Networks: The Definitive Guide* (Sebastopol, California: O'Reilly & Associates, 2002), 199.

it receives a synchronization value greater than its current value. Stations entering a power-save mode may inform a PC through the frame control field of a message. The AP will then buffer transmissions to the station. A station is informed that it has buffered transmissions waiting when it wakes periodically to receive beacon frames. It can then request transmission. A station in active mode can receive frames at any time during a contention-free period. A station in a power-save mode will periodically enter the active mode to receive beacons, broadcast, multicast, and buffered data frames.[9]

Medium Access Control (MAC) Concepts and Architecture

The IEEE 802.11 MAC is common to all IEEE 802.11 PHY layers and specifies the functions and protocols required for control and access. The MAC layer is responsible for managing data transfer from higher-level functions to the physical media. Figure 2-4 illustrates this relationship to the *Open Systems Interconnection* (OSI) model.

MAC Layer Services

Devices using the IEEE 802.11 PHY and MAC as part of a WLAN are called *stations*. Stations can be endpoints or APs. APs are stations that act as part of the DS and facilitate the distribution of data between endpoints. The MAC provides nine logical services: authentication, deauthentication, association, disassociation, reassociation, distribution, integration, privacy, and data delivery. An AP uses all nine services. An endpoint uses authentication, deauthentication, privacy, and data delivery. Each service utilizes a set of messages with information elements that are pertinent to the services. Table 2-2 describes these services.

---

[9]Ibid., 128.

Figure 2-4
IEEE 802.11
standards
mapped to the
OSI reference
model

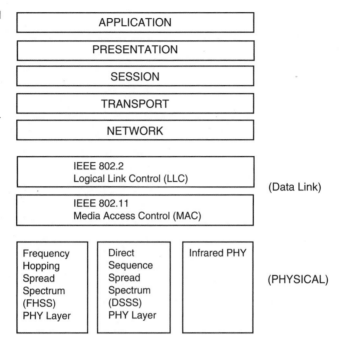

MAC Layer Architecture

As illustrated in Figure 2-4, both the PHY and MAC layers are conceptually divided into management and data transfer capabilities. The PHY management capability is provided by the *PHY layer management entity* (PLME). The MAC management capability is provided by the *MAC layer management entity* (MLME). The PLME and the MLME exchange information about PHY medium capabilities through a *Management Information Base* (MIB) (see the following paragraphs for more information). This is a database of physical characteristics such as possible transmission rates, power levels, and antenna types. Some of these characteristics are static and some can be changed by a management entity. These management functions support the main purpose of the MAC, which is to transfer data elements. These data elements originate in the *Logical Link Control* (LLC) layer. Packages of data passed to the MAC from the LLC are called *MAC service data units* (Medusa). In order to transfer the Medusa to the PHY, the MAC uses messages (frames) containing

Table 2-2 IEEE 802.11 MAC Services and Agents

MAC Service	Definition	Station Type
Authentication	Because WLANs have limited physical security to prevent unauthorized access, 802.11 defines the authentication services needed to control access to the WLAN. The goal of the authentication service is to provide access control equal to a wired LAN. The authentication service provides a mechanism for one station to identify another station. Without this proof of identity, the station is not allowed to use the WLAN for data delivery. All 802.11 stations, whether they are part of an IBSS or *extended service set* (ESS) network, must use the authentication service prior to communicating with another station.	Endpoint and AP
Open system authentication	This is the default authentication method, which is a very simple, two-step process. First, the station wanting to authenticate with another station sends an authentication management frame containing the sending station's identity. The receiving station then sends back a frame alerting whether it recognizes the identity of the authenticating station.	
Shared key authentication	This type of authentication assumes that each station has received a secret shared key through a secure channel independent of the 802.11 network. Stations authenticate through shared knowledge of the secret key. Use of shared key authentication requires the implementation of encryption via the *Wired Equivalent Privacy* (WEP) algorithm.	
Deauthentication	This type removes an existing authentication. The deauthentication service is used to eliminate a previously authorized user from any further use of the network. Once a station is deauthenticated, that station is no longer able to access the WLAN without performing the authentication function again. Deauthentication is a notification and cannot be refused. For example, when a station wants to be removed from a BSS, it can send a deauthentication management frame to the associated AP to notify the AP of the removal from the network. An AP can also deauthenticate a station by sending a deauthentication frame to the station.	Endpoint and AP

Table 2-2   IEEE 802.11 MAC Services and Agents (continued)

MAC Service	Definition	Station Type
Association	Association maps a station to an AP and enables the AP to distribute data to and from the station. The association service is used to make a logical connection between a mobile station and an AP. Each station must become associated with an AP before it is allowed to send data through the AP onto the DS. The connection is necessary in order for the DS to know where and how to deliver data to the mobile station. The mobile station invokes the association service once and only once, typically when the station enters the BSS. Each station can associate with one AP, although an AP can associate with multiple stations.	AP
Disassociation	This breaks an existing association relationship. The disassociation service is used either to force a mobile station to eliminate an association with an AP or for a mobile station to inform an AP that it no longer requires the services of the DS. When a station becomes disassociated, it must begin a new association to communicate with an AP again. An AP may force a station or stations to disassociate because of resource restraints; the AP is shutting down or being removed from the network for a variety of reasons. When a mobile station is aware that it will no longer require the services of an AP, it may invoke the disassociation service to notify the AP that the logical connection to the services of the AP from this mobile station is no longer required. Stations should disassociate when they leave a network, although nothing in the architecture ensures this will happen. Disassociation is a notification and can be invoked by either associated party. Neither party can refuse the termination of the association.	AP
Reassociation	This type transfers an association between APs. Reassociation enables a station to change its current association with an AP. The reassociation service is similar to the association service, with the exception that it includes information about the AP with which a mobile station has been previously associated. A mobile station will use the reassociation service repeatedly as it moves throughout the ESS, loses contact with the AP with which it is associated, and needs to become associated with a new AP.	

Table 2-2 IEEE 802.11 MAC Services and Agents (continued)

MAC Service	Definition	Station Type
	By using the reassociation service, a mobile station provides information to the AP to which it will be associated and information pertaining to the AP to which it will be disassociated. This enables the newly associated AP to contact the previously associated AP to obtain frames that may be waiting there for delivery to the mobile station as well as other information that may be relevant to the new association. The mobile station always initiates reassociation.	AP
Privacy	This type prevents the unauthorized viewing of data through the use of the WEP algorithm. The privacy service of IEEE 802.11 is designed to provide an equivalent level of protection for data on the WLAN as that provided by a wired network with restricted physical access. This service protects that data only as it traverses the wireless medium. It is not designed to provide complete protection of data between applications running over a mixed network. With a wireless network, all stations and other devices can hear data traffic taking place within range on the network, seriously impacting the security level of a wireless link. IEEE 802.11 counters this problem by offering a privacy service option that raises the security of the 802.11 network to that of a wired network. The privacy service, applying to all data frames and some authentication management frames, is an encryption algorithm based on 802.11.	Endpoint and AP
Distribution	This authentication provides data transfer between stations through the DS. Distribution is the primary service used by an 802.11 station. A station uses the distribution service every time it sends MAC frames across the DS. The distribution service provides the distribution with only enough information to determine the proper destination BSS for the MAC frame. The three association services (association, reassociation, and disassociation) provide the necessary information for the distribution service to operate. Distribution within the DS does not necessarily involve any additional features outside of the association services, although a station must be associated with an AP for the distribution service to forward frames properly.	AP

Table 2-2 IEEE 802.11 MAC Services and Agents (continued)

MAC Service	Definition	Station Type
Data delivery	This provides data transfer between stations.	Endpoint and AP
Integration	This provides data transfer between the DS of an IEEE 802.11 LAN and a non-IEEE 802.11 LAN. The station providing this function is called a *portal*. The integration service connects the 802.11 WLAN to other LANs, including one or more wired LANs or 802.11 walls. A portal performs the integration service. A portal is an abstract architectural concept that typically resides in an AP, although it could be part of a separate network component entirely. The integration service translates 802.11 frames to frames that can traverse another network.	AP

Source: Intelligraphics and LaRocca, 135

functionality-related fields. Three types of MAC frames are available: control, management, and data. One of these messages is called a *MAC protocol data unit* (MPDU). The MAC passes MSDU to the PHY layer through the *Physical Layer Convergence Protocol* (PLCP). The PLCP is responsible for translating Medusa into a format that is *physical medium dependent* (PMD). The PMD layer transfers the data onto the medium.

MAC data transfer is controlled through two distinct coordination functions. The first is the *distributed coordination function* (DCF), which defines how users contend for the medium as peers. DCF data transfers are not time sensitive and delivery is asynchronous. The second is the *point coordination function* (PCF), which provides centralized traffic management for data transfers that are sensitive to delay and require contention-free access.[10]

Management Information Base (MIB)   802.11 contains extensive management functions to make the wireless connection appear much like a regular wired connection. The complexity of the additional management functions results in a complex management entity with dozens of variables. For ease of use, the variables have been organized into an MIB so that network managers can benefit from taking a structured view of the 802.11 parameters. The formal specification of the 802.11 MIB is Annex D of the 802.11 specification. The 802.11 MIB is designed by the 802.11 Working Group.[11]

Distributed Coordination Function (DCF)   The DCF defines how the medium is shared among members of the wireless network. It provides mechanisms for negotiating access to the wireless medium as well as mechanisms for reliable data delivery. One of the fundamental differences between wired and wireless media is that it is difficult to detect and manage data collisions on wireless media.

---

[10]LaRocca, 134–135.

[11]Gast, 383.

The primary reason for this is that stations in a radio network are not guaranteed to hear every other station's transmissions. This is typically the case when an AP is used in IEEE 802.11's infrastructure BSS and is called the *hidden-node problem.*

Point Coordination Function (PCF)   The PCF polls associated stations and manages frame transmissions on their behalf. A station performing PCF traffic management is called a *point coordinator* (PC). The PCF is an optional capability that provides connection-oriented services for delay-sensitive traffic. The PCF is more complex to implement, but it provides a moderate level of priority frame delivery for time-sensitive transmissions.

The PC uses beacon signals to broadcast for the duration of a contention-free period to all associated stations. This causes them to update their *network allocation vector* (NAV) and wait for the duration of the contention-free period. In addition, stations must wait for the *PCF interframe space* (PIFS) interval to further decrease the possibility of data collisions. The transmission of the additional polling and ACK messages required by the PCF is optimized through piggybacking multiple messages in a single transmission. For example, the PC may append both ACKs of previous transmissions and polling messages for new traffic to a data frame. This enables the transmission to avoid waiting for the interframe interval specified for individual frame transmissions.[12]

The basic access method for 802.11 is the DCF, which uses CSMA/CA. This requires each station to listen for other users. If the channel is idle, the station may transmit. If the station is busy, it waits until transmission stops and then enters into a random back-off procedure. This prevents multiple stations from seizing the medium immediately after completing the preceding transmission.

Packet reception in DCF requires acknowledgement, as shown in Figure 2-5. The period between the completion of packet transmission and the start of the ACK frame is one *short interframe space*

---

[12]LaRocca, 140–141.

Figure 2-5
CSMA/CA back-
off algorithm

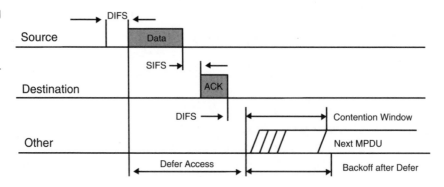

(SIFS). ACK frames have a higher priority than other traffic. Fast acknowledgement is one of the salient features of the 802.11 standard, because it requires ACKs to be handled at the MAC sublayer.

Transmissions other than ACKs must wait at least one *DCF inter-frame space* (DIFS) before transmitting data. If a transmitter senses a busy medium, it determines a random back-off period by setting an internal timer to an integer number of slot times. Upon expiration of the DIFS, the timer begins to decrement. If the time reaches zero, the station may begin transmission. If the channel is seized by another station before the timer reaches zero, the timer setting is retained at the decremented value for subsequent transmission. This method relies on the physical carrier sense. The underlying assumption is that every station can hear all the other stations.[13]

802.11 Architecture

IEEE 802.11 supports three basic topologies for WLANs: the IBSS, BSS, and ESS. All three configurations are supported by the MAC layer implementation.

The 802.11 standard defines two modes: ad hoc/IBSS and infrastructure mode. Logically, an *ad hoc configuration* is analogous to a

---

[13]Zyren and Petrick.

peer-to-peer office network in which no single node is required to function as a server (see Figure 2-6). IBSS WLANs include a number of nodes or wireless stations that communicate directly with one another on an ad hoc, peer-to-peer basis, building a full-mesh or partial-mesh topology. Generally, ad hoc implementations cover a limited area and are not connected to any larger network.

Using *infrastructure mode*, the wireless network consists of at least one AP connected to the wired network infrastructure and a set of wireless end stations. This configuration is a BSS (see Figure 2-7). Since most corporate WLANs require access to the wired LAN for services (file servers, printers, and Internet links), they will operate in infrastructure mode and rely on an AP that acts as the logical server for a single WLAN cell or channel. Communications between two nodes, A and B, actually flow from node A to the AP and then from the AP to node B. The AP is necessary to perform a bridging function and connect multiple WLAN cells or channels as well as connect WLAN cells to a wired enterprise LAN.

An ESS is a set of two or more BSSs forming a single subnetwork. ESS configurations consist of multiple BSS cells that can be linked by either wired or wireless backbones. IEEE 802.11 supports ESS configurations in which multiple cells use the same channel and use different channels to boost aggregate throughput (see Figure 2-8).

Figure 2-6

Wireless ad hoc network

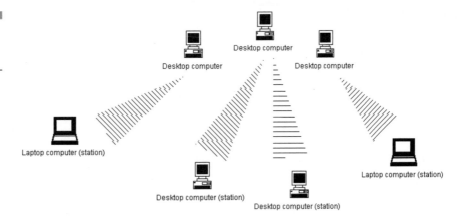

Figure 2-7
Wireless BSS

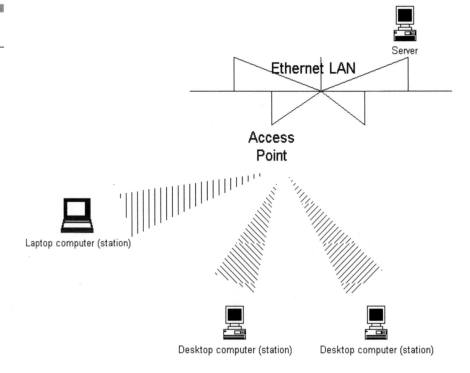

Figure 2-8
802.11 ESS

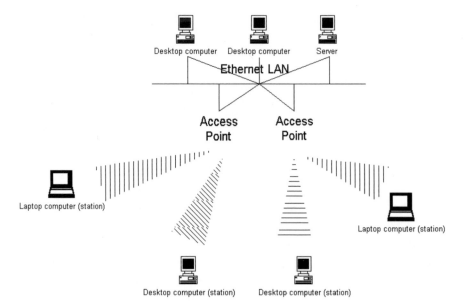

802.11 Components

802.11 defines two pieces of equipment, a wireless station, which is usually a PC equipped with a wireless NIC, and an AP, which acts as a bridge between the wireless and wired networks. An AP usually consists of a radio, a wired network interface (802.3, for example), and bridging software conforming to the 802.11d bridging standard. The AP acts as the base station for the wireless network, aggregating access for multiple wireless stations onto the wired network. Wireless end stations can be 802.11 PC card, *Peripheral Component Interconnection* (PCI), or *Industry Standard Architecture* (ISA) NICs, or embedded solutions in non-PC clients (such as an 802.11-based telephone handset).

An 802.11 WLAN is based on a cellular architecture. Each cell (BSS) is connected to the base station or AP. All APs are connected to a DS, which is similar to a backbone, usually Ethernet or wireless. All mentioned components appear as an 802 system for the upper layers of OSI and are known as the ESS.

The 802.11 standard does not constrain the composition of the DS; therefore, it may be 802 compliant or nonstandard. If data frames need to transmit to and from a non-IEEE 802.11 LAN, then these frames, as defined by the 802.11 standard, enter and exit through a portal. The portal provides logical integration between existing wired LANs and 802.11 LANs.

When the DS is constructed with 802-type components, such as 802.3 (Ethernet) or 802.5 (Token Ring), then the portal and the AP are the same, acting as a translation bridge. The 802.11 standard defines the DS as an element that interconnects BSSs within the ESS via APs. The DS supports the 802.11 mobility types by providing the logical services necessary to handle address-to-destination mapping and the seamless integration of multiple BSSs. An AP is an addressable station, providing an interface to the DS for stations located within various BSSs. The IBSS and ESS networks are transparent to the LLC layer.[14]

---

[14]Plamen Nedeltchev, "WLANs and the 802.11 Standard," a white paper from Cisco Systems, www.cisco.com/warp/public/784/packet/jul01/pdfs/whitepaper.pdf

Mobility

The mobility of wireless stations may be the most important feature of a WLAN. The chief motivation of deploying a WLAN is to enable stations to move about freely from location to location either within a specific WLAN or between different WLAN segments.

For compatibility purposes, the 802.11 MAC must appear to the upper layers of the network as a standard 802 LAN. The 802.11 MAC layer is forced to handle station mobility in a fashion that is transparent to the upper layers of the 802 LAN stack. This forces functionality into the 802.11 MAC layer that is typically handled by upper layers in the OSI model.[15]

To understand this design restriction, it is first important to appreciate the difference between true mobility and mere portability. Portability certainly results in a net productivity gain because users can access information resources wherever it is convenient to do so. At the core, however, portability removes only the physical barriers to connectivity. It is easy to carry a laptop between several locations, so people do. However, portability does not change the ritual of connecting to networks at each new location. It is still necessary to physically connect to the network and reestablish network connections, and network connections cannot be used while the device is being moved.

Mobility removes further barriers, most of which are based on the logical network architecture. Network connections stay active even while the device is in motion. This is critical for tasks requiring persistent, long-lived connections, which may be found in database applications.

802.11 is implemented at the link layer and provides link-layer mobility. The *Internet Protocol* (IP) does not allow this. 802.11 hosts can move within the last network freely, but IP, as it is currently deployed, provides no way to move across subnet boundaries. To the IP-based hosts of the outside world, the *virtual private network* (VPN)/access control boxes are the last-hop routers. To access an 802.11 wireless station with an IP address on the wireless network,

---

[15]"Introduction to IEEE 802.11," a white paper from Intelligraphics, www.intelligraphics.com/articles/80211_article.html.

it is possible to simply go through the IP router to the target network regardless of whether a wireless station is connected to the first or third AP. The target network is reachable through the last-hop router. As far as the outside world can tell, the wireless station might as well be a workstation connected to an Ethernet.

A second requirement for mobility is that the IP address does not change when connecting to any of the APs. New IP addresses interrupt open connections. If a wireless station connects to the first AP, it must keep the same address when it connects to the third AP.

A corollary to the second requirement is that all the wireless stations must be on the same IP subnet. As long as a station stays on the same IP subnet, it does not need to reinitialize its networking stack and can keep its *Transmission Control Protocol* (TCP) connections open. If it leaves the subnet, though, it needs to get a new IP address and reestablish any open connections. Multiple subnets are not forbidden, but if you have different IP subnets, seamless mobility between subnets is not possible.

The single IP subnet backbone restriction is a reflection on the technology deployed within most organizations. Mobile IP was standardized in late 1996 in RFC 2002, but it has yet to see widespread deployment. Until Mobile IP can be deployed, network designers must live within the limitations of IP and design networks based on fixed locations for IP addresses. A backbone network may be physically large, but it is fundamentally constrained by the requirement that all APs connect directly to the backbone router (and each other) at the link layer.[16]

Conclusion

Some may argue that Morse code and the telegraph were the first technologies that transmitted data via the airwaves (dots and dashes versus ones and zeros). The ability to transmit data over the airwaves presents some exciting opportunities for business net-

---

[16]Gast, 295–296.

works. Businesses worldwide have made the switch from wired to wireless in order to save money and increase employee productivity.

IEEE 802.11b is a subvariant of 802.11, which is a standard that digresses slightly from the OSI model in that it provides a standard for wireless data transmission. To do this, the standard defines the MAC and PHY layers of the OSI model for use of DSSS (for 802.11b). The MAC layer is responsible for managing data transfer from higher-level functions to PHY media. This standard details how data is modulated for transmission and correlated at the receiving end. The topology of wireless networks is fairly simple. In a BSS, an AP is connected to an existing LAN from which wireless stations can access the network. An ESS extends this topolgy to expand the network. Using an ad hoc topology, stations (PCs) can communicate directly with one another. Mobility measures permit wireless users to access the wireless network from any point on the network and maintain their connection regardless of where they roam on the network.

Range Is Not an Issue

Range Is a Matter of Engineering

One of the major misperceptions regarding 802.11b and other wireless protocols is that the range is limited to 100 meters and thus proves impractical as a last-mile solution. The truth is that with proper engineering, 802.11b can reach beyond 20 miles point to point. In the quest for *Public Switched Telephone Network* (PSTN) bypass, this is one of the most exciting revelations. By steering an antenna in the direction of the subscriber's home, the service provider can bring broadband wireless to masses of homes without stringing a single strand of copper wire, digging up a single street, or engaging in a single legal battle for right-of-way.

Furthermore, new wireless protocols for *metro area networks* (MANs) provide for the construction of wireless networks that can cover entire cities. Ad hoc peer-to-peer networks stretch the range of a wireless network with a minimum of investment.

This chapter covers the science of antennas and how proper engineering can stretch the most modest resources to deliver essential services to the home. Secondly, this chapter explains how 802.11b antenna systems can be used to stretch the range of delivery out to a number of miles to blanket large metropolitan areas and even reach out to rural subscribers. The most important part in designing a broadband wireless network is the inclusion of the new protocol 802.16 in the deployment of *wireless metro area networks* (WMANs) to feed suburban 802.11b networks. Other technologies such as mesh networks also extend the range of broadband wireless networks.

In data networking, the success of 802.11 has inexorably linked it with *radio frequency* (RF) engineering. Whereas a wired network requires little or no knowledge on the part of the installer of how data travels via an Ethernet cable, a wireless network requires a strong knowledge of radios and antennas. The following paragraphs provide a basic overview of wireless transmission systems.

RF Components

RF systems complement wired networks by extending them. Different components may be used depending on the frequency and the distance that signals are required to reach, but all systems are fundamentally the same and made from a relatively small number of components. Three RF components of particular interest to 802.11 users are antennas, sensitive receivers, and amplifiers. Antennas are of general interest since they are the most tangible feature of an RF system.

Antennas

Antennas are the most critical component of any RF system as they convert electrical signals on wires into radio waves, and vice versa. To function at all, an antenna must be made of conducting material. Radio waves hitting an antenna cause electrons to flow in the conductor and create a current. Similarly, applying a current to an antenna creates an electric field around the antenna. The electric field changes as the current to the antenna changes. A changing electric field causes a magnetic field, and the wave is off.

The size of the antenna used depends on the frequency; the higher the frequency, the smaller the antenna. The shortest simple antenna possible at any frequency is one-half wavelength long. This rule of thumb accounts for the huge size of radio broadcast antennas and the small size of mobile phones. An AM station broadcasting at 830 kHz at a wavelength of about 360 meters has a correspondingly large antenna, but an 802.11b network interface operating in the 2.4 GHz band has a wavelength of just 12.5 centimeters. With some engineering tricks, an antenna can be incorporated into a PC card or the top of a laptop computer.

Antennas can also be designed with directional preference. Many antennas are omnidirectional, which means they send and receive signals from any direction. Some applications may benefit from directional antennas, which radiate and receive on a narrower portion of the field. Figure 3-1 compares the radiated power of omnidirectional and directional antennas.

For a given amount of input power, a directional antenna can reach farther with a clearer signal. The antenna must also have much higher sensitivity to radio signals in the dominant direction. When wireless links are used to replace wireline networks, directional antennas are often used. Mobile telephone network operators

Figure 3-1
Radiated power
and reach of
antennas:
omnidirectional
and directional

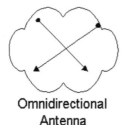

Omnidirectional
Antenna

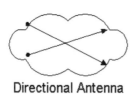

Directional Antenna

also use directional antennas when cells are subdivided. 802.11 networks typically use omnidirectional antennas for both ends of the connection.

Antennas are the most likely to be separated from the rest of the electronics. A transmission line (some kind of cable) between the antenna and the transceiver is also necessary. Transmission lines usually have an impedance of 50 ohms. In terms of practical antennas for 802.11 devices in the 2.4 GHz band, the typical wireless PC card has a built-in antenna. The antenna plugs into the card.

Wireless cards all have built-in antennas, but these antennas are, at best, minimally adequate. If you are planning to cover an office or an even larger area, such as a campus, you will almost certainly want to use external antennas for your *access points* (APs). When considering specialized antennas, you need to pay attention to the following specifications.

Antenna Type The antenna type determines its radiation pattern: omnidirectional, bidirectional, or unidirectional. Omnidirectional antennas are good for covering large areas; bidirectional antennas are particularly good for covering corridors; and unidirectional antennas are the best for setting up point-to-point links between buildings or even different sites. It usually follows that the higher the gain, the narrower the beam.

Factors Affecting Range   It's tempting to think that you can put up a high-gain antenna and a power amplifier and cover a huge territory, thus economizing on APs and serving a large number of users at once. This is not a particularly good idea. The larger the area you cover, the more users your APs must serve. A good upper bound to aim for is 20 to 30 users per wireless card per AP. A single AP covering a large territory may look like a good idea, and it may even work well while the number of users remains small. However, if a network is successful, the number of users will grow quickly and the network will soon exceed the AP's capacity.[1] Once this happens, it is necessary to install more APs and divide the original cell into several smaller ones and lower the power output at all of the cells.

Sensitive Receivers

Besides the antenna, the most critical item in a Wi-Fi system is the receiver. In particular, it is important to look for the receiver sensitivity. The receiver sensitivity is the lowest level signal that can be decoded by the receiver. The lower the receiver sensitivity, the longer the range.

Amplifiers

Amplifiers make signals larger. Signal boost, or gain, is measured in *decibels* (dB). Amplifiers can be broadly classified into three categories: low noise, high power, and everything else. *Low-noise amplifiers* (LNAs) are usually connected to an antenna to boost the received signal to a level that is recognizable by the electronics to which the RF system is connected. LNAs are also rated with a noise figure, which is the measure of how much extraneous information the amplifier introduces to the *signal-to-noise ratio* (SNR). Smaller

---

[1]Matthew Gast, *802.11b Wireless Networks: The Definitive Guide* (Sebastopol, California: O'Reilly & Associates, 2002), 316–322.

noise figures enable the receiver to hear smaller signals and thus provide a greater range.

High-power amplifiers (HPAs) are used to boost a signal to the maximum power possible before transmission. Output power is measured in dBm, which are related to watts. Amplifiers are subject to the laws of thermodynamics; they give off heat in addition to amplifying the signal. The transmitter in an 802.11 PC card is necessarily low power because it needs to run off a battery if it is installed in a laptop, but it is possible to install an external amplifier at feed APs. This amplifier can be connected to the power grid where power is more plentiful. This is where things can get tricky with respect to compliance with regulations. 802.11 devices are limited to 1 watt of power output and 4 watts of *effective radiated power* (ERP). ERP multiplies the transmitter's power output by the gain of the antenna minus the loss in the transmission line. With a 1 watt amplifier, an antenna that provides 8 dB of gain, and 2 dB of transmission line loss, the result is an ERP of 4 watts; the total system gain is 6 dB, which multiplies the transmitter's power by a factor of 4.

802.11b at 20 to 72 Miles

It is possible to have point-to-multipoint links in excess of over 1,500 feet with ordinary equipment at the client side. Using high-gain antennas, sensitive receivers, and amplifiers, if necessary, it is possible to achieve Ethernet-like speeds over point-to-point links that exceed 20 miles. An experiment proved that it is theoretically possible to drive 802.11b signals well over 20 miles using stock equipment.[2] In fact, a 72-mile link from San Diego to San Clemente Island has been established by Hans Werner-Braun using some specialized 802.11 equipment on the 2.4 GHz band.[3] Figure 3-2 illustrates the range of 802.11b.

---

[2]Rob Flickinger, "A Wireless Long Shot," O'Reilly Network, www.oreillynet.com/pub/a/wireless/2001/05/03/longshot.html, May 3, 2001.

[3]Bob Brewin, "San Diego Wireless Net Installs 72-Mile, 2.4-GHz Link," Computer World, www.computerworld.com/mobiletopics/mobile/story/0,10801,75830,00.html, November 12, 2002.

In summary, 802.11b, by itself, is *not* limited to a range of 100 meters. Its maximum range is in excess of 20 miles. A scatological comparison is that of the telephone central office, where the maximum range of the signal over copper wire is 18,000 feet (3 miles) without a repeater. It could be argued that the maximum, unboosted range of 802.11b exceeds that of the PSTN.

Architecture: The Large Network Solution

Although a point-to-point 802.11b connection may have a range of 20 miles and a point-to-multipoint connection may have somewhat less range, building a wireless network to compete with the PSTN is considerably more complicated. Issues revolving around bandwidth sharing and frequency contention require a multitiered strategy for building a WMAN to replace the PSTN in a given municipality.

Overcoming limitations of range can be achieved through properly planning the architecture of a wireless network. Four elements of network architecture can be employed to extend the maximum range of 802.11b and its associated wireless protocols to cover an entire metropolitan area. First, a WMAN is fed from an *Internet Protocol* (IP) backbone at a high bandwidth—say, 100 Mbps. This WMAN would operate at a licensed frequency to ensure a high quality of transmission devoid of interference. The chief subscribers of the WMAN would be *wireless Internet service providers* (WISPs). The

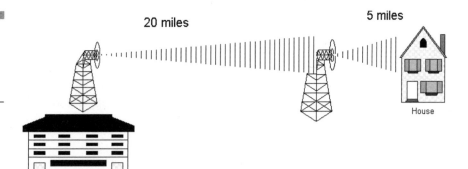

Figure 3-2
The range of 802.11b exceeds 20 miles.

20 miles

5 miles

House

Office or WISP

WMAN would then feed lesser networks, such as *wireless wide area networks* (WWANs). The WWANs could operate at 802.11a bandwidth (54 Mbps) at a frequency in the 5.8 GHz range. Subscribers of the WWAN would include large enterprises and smaller WISPs. The WWAN would, in turn, feed *wireless local area networks* (WLANs). WLANs would feed residences and small businesses. *Wireless personal area networks* (WPANs) would feed off WLANs to serve components within a given residence. Finally, an ad hoc peer-to-peer network, consisting of subscriber devices, intelligent APs, and wireless routers, could extend the network even further with little infrastructure cost. Figure 3-3 illustrates this process.

Metro Area Networks (MANs)

The WMAN encompasses a range of radio- and laser-based technologies targeted at providing wireless networking over distances of a few hundred meters to several miles. Wireless broadband, *broadband wireless access* (BWA), *wireless local loop* (WLL), fixed wireless, and wireless cable all refer to technologies that can be used to deliver telecommunications services over the last few miles of the network. Wireless broadband and BWA are general terms referring to high-speed wireless networking systems. WLL is derived from the wired telephony term *local loop*, which refers to the connection between a local telephone switch and a subscriber. WLL and fixed wireless generally refer to the delivery of voice and data services between fixed locations over a high-speed wireless medium. Some new market entrants offer mobile applications of this technology. Fixed wireless includes *Local Multipoint Distribution Service* (LMDS), *Multichannel Multipoint Distribution Service* (MMDS), *Unlicensed National Information Infrastructure* (U-NII) systems, and similar networks. Wireless cable usually refers to MMDS systems used to deliver television signals such as the *Instructional Television Fixed Service* (ITFS).

Two basic network topologies are supported by these systems. The simplest is a point-to-point system providing a high-speed wireless connection between two fixed locations. Bandwidth is not shared, but links typically require line of sight between the two antennas. The

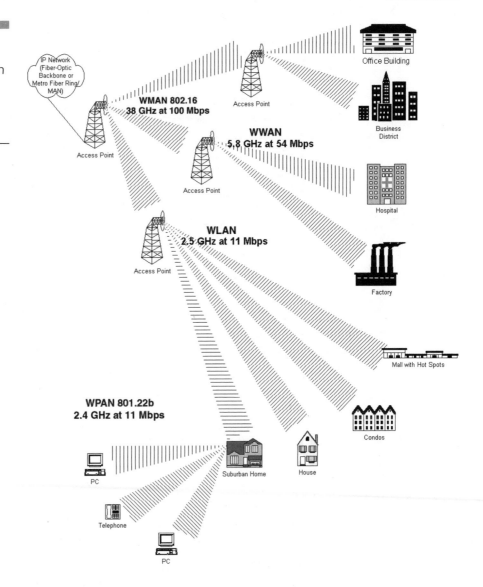

■■■ ■■■ ■■■
Figure 3-3
Covering a
metro area with
WMANs,
WWANs,
WLANs, and
WPANs

second topology is a point-to-multipoint network where a signal is broadcast over an area (called a *cell*) and communicates with fixed subscriber antennas in the cell. Because bandwidth in the cell is finite and is shared among all users, performance may be a concern in high-density cells. Systems of different frequencies may be combined to cover an area where terrain or other obstructions prevent full coverage.

Other than frequency, the main difference between fixed wireless systems, and cellular, WLAN, and WPAN networks is the mobility of subscriber equipment. There has been some discussion about adding support for mobile subscriber equipment to fixed wireless systems. The addition of mobility support would enable these BWA systems to potentially function as *fourth-generation* (4G) cellular networks, delivering subscriber speeds of several megabits. Several technical, regulatory, and commercial hurdles must still be overcome before this could become a reality, but companies such as Wi-Fi have already started examining products targeted at this potential application.[4]

Local Multipoint Distribution Service (LMDS)

LMDS is a fixed wireless, radio-based technology. In North America, LMDS operates in the 28 to 31 GHz frequency range, but it may operate anywhere from 2 to 40 GHz in other regions. In 1998, the *Federal Communications Commission* (FCC) held an auction for this spectrum, dividing each geographical area into the A Block and B Block. The A Block had a bandwidth of 1.5 GHz and the B Block had a bandwidth of 150 MHz. The intent was for the auction winners to deploy high-speed voice and data communications services in the last mile. The realities of deployment have not yet lived up to that vision.

The network topology of LMDS uses a central transmitter that sends its signal over a cell with a radius of 5 km or less. Antennas are usually placed on rooftops for line of sight to the central transmitter. This is because *first-generation* (1G) LMDS equipment uses radio technology that is affected by hills, walls, trees, and other physical barriers. This limitation may be lessened as equipment starts to adopt more advanced spectrum utilization techniques such as *orthogonal frequency division multiplexing* (OFDM).

---

[4]James and Ruth LaRocca, *802.11 Demystified* (New York: McGraw-Hill, 2002), 55–56.

As a high-frequency outdoor radio technology, LMDS performance and range vary depending on weather conditions. LMDS has a range of less than 5 km and supports gigabit speeds, although services are usually offered at a much lower rate. The physics of the 30 GHz signal make it about a millimeter in length; this spectrum is sometimes referred to as *millimeter wave spectrum*. One effect of having such a small wavelength is that rain can effectively block the signal. In areas where rain is a factor, a lower frequency is required. A higher frequency allows faster data rates, but it also limits range, requiring more equipment to cover the same area as a lower-frequency technology. LMDS bandwidth in a specific area is shared among all the users like cable. To ensure end-user performance, networks must be built with enough capacity to handle sporadic peak loads and unexpected growth in the subscriber base. In addition, there are no standards governing LMDS implementations, leading to a number of incompatible proprietary solutions. Higher network deployment costs make 1G LMDS networks more suitable for high-margin business applications rather than residential use.[5]

802.16: Protocol for WMANs

An 802.16 wireless service provides a communications path between a subscriber site and a core network (the network to which 802.16 is providing access). Examples of a core network are the public telephone network and the Internet. IEEE 802.16 standards are concerned with the air interface between a subscriber's transceiver station and a base transceiver station.

Protocols defined specifically for wireless transmission address issues related to the transmission of blocks of data over a network. The standards are organized into a three-layer architecture. The lowest layer, the physical layer, specifies the frequency band, the modulation scheme, error-correction techniques, synchronization

---

[5]Ibid., 58–59.

between transmitter and receiver, the data rate, and the *time-division multiplexing* (TDM) structure.[6]

IEEE 802.16 addresses first-mile applications of wireless technology to link commercial and residential buildings to high-rate core networks and thereby provide access to those networks. The 802.16 group's work has been primarily aimed at a point-to-multipoint topology with a cellular deployment of base stations, which are each tied to core networks and in contact with fixed-wireless subscriber stations.

Working Group 802.16 is now completing a draft of the IEEE 802.16 Standard Air Interface for Fixed Broadband Wireless Access Systems. The document includes a flexible *Media Access Control* (MAC) layer. The accompanying *physical* (PHY) layer is designed for 10 to 66 GHz, informally known as the LMDS spectrum. The standard is not yet final, but the draft is stable and has passed the Working Group's letter ballot, pending resolution of comments proposed to improve it. Publication is planned for late this year.[7]

For transmission from subscribers to a base station, the standard uses the *Demand Assignment Multiple Access—Time Division Multiple Access* (DAMA—TDMA) technique. DAMA is a capacity assignment technique that adapts as needed to respond to demand changes among multiple stations. TDMA is the technique of dividing time on a channel into a sequence of frames, each consisting of a number of slots, and allocating one or more slots per frame to form a logical channel.

With DAMA-TDMA, the assignment of slots to channels varies dynamically. For transmission from a base station to subscribers, the standard specifies two modes of operation: one targeted to support a continuous transmission stream (mode A), such as audio or video, and one targeted to support a burst transmission stream (mode B), such as IP-based traffic. Both are TDM schemes.

Above the physical layer are the functions associated with providing service to subscribers. These functions include transmitting data

[6]William Stallings, "IEEE 802.16 for Broadband Wireless," Network World, www.nwfusion.com/news/tech/2001/0903tech.html, September 3, 2001.

[7]Roger Marks, "Broadband Access: IEEE Takes on Broadband Wireless," EE Times, www.eetimes.com/story/OEG20010606S0008, January 4, 2002.

in frames and controlling access to the shared wireless medium, and are grouped into the MAC layer. The MAC protocol defines how and when a base station or subscriber station may initiate transmission on the channel. Because some of the layers above the MAC layer, such as *Asynchronous Transfer Mode* (ATM), require *quality of service* (QoS), the MAC protocol must be able to allocate radio channel capacity to satisfy service demands.

In the downstream direction (base station to subscriber stations), only one transmitter is available and the MAC protocol is relatively simple. In the upstream direction, multiple subscriber stations compete for access, resulting in a more complex MAC protocol. In both directions, a TDMA technique is used in which the data stream is divided into a number of time slots.

The sequence of time slots across multiple TDMA frames that are dedicated to one subscriber forms a logical channel, and MAC frames are transmitted over that logical channel. IEEE 801.16.1 is intended to support individual channel data rates from 2 to 155 Mbps.

Above the MAC layer is a convergence layer that provides functions specific to the service being provided. For IEEE 802.16.1, bearer services include digital audio/video multicast, digital telephony, ATM, Internet access, wireless trunks in telephone networks, and frame relay. Figure 3-4 depicts how the 802.16 protocol works for WMANs.[8]

Consecutive Point Network (CPN)

In a WMAN, the reliability of the network can be ensured by implementing *consecutive point network* (CPN) technology. Like a *Synchronous Optical Network* (SONET) fiber ring, the data flow of the network around the wireless ring would reverse flow in the event of a disruption in the network (see Figure 3-5). This ensures that only a limited part of the network is down due to a disruption.

---

[8]William Stallings, IEEE 802.16 for Broadband Wireless, *Network World*, www.nwfusion.com/news/tech/2001/0903tech.html, September 3, 2001.

Figure 3-4
A WMAN

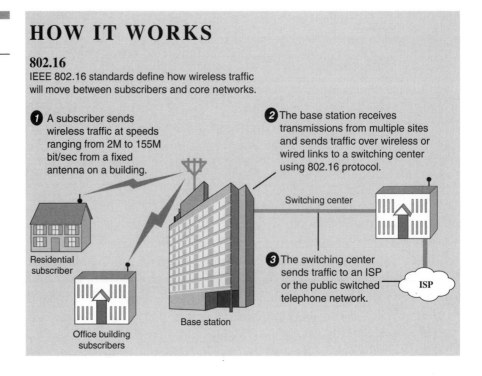

HOW IT WORKS

802.16
IEEE 802.16 standards define how wireless traffic will move between subscribers and core networks.

❶ A subscriber sends wireless traffic at speeds ranging from 2M to 155M bit/sec from a fixed antenna on a building.

❷ The base station receives transmissions from multiple sites and sends traffic over wireless or wired links to a switching center using 802.16 protocol.

Switching center

Residential subscriber

❸ The switching center sends traffic to an ISP or the public switched telephone network.

ISP

Office building subscribers

Base station

Extending Range Via an Ad Hoc Peer-to-Peer Network

Ad hoc peer-to-peer technologies extend the maximum range of Wi-Fi networks from distances typically measured in hundreds of feet to several miles. The product adds multihopping peer-to-peer capabilities to off-the-shelf 802.11 cards.

Software is utilized to turn WLAN cards into router repeaters. The result is a system that enables users who are out of range of an AP to hop through one or more other nearby users until they connect to the AP. The software also automatically routes transmissions from congested APs to uncongested ones. Overall network performance is enhanced in addition to the dramatic increases in effective range. In

Figure 3-5
CPNs—note that like a SONET ring, data flow reverses itself in case of disruption in the network.

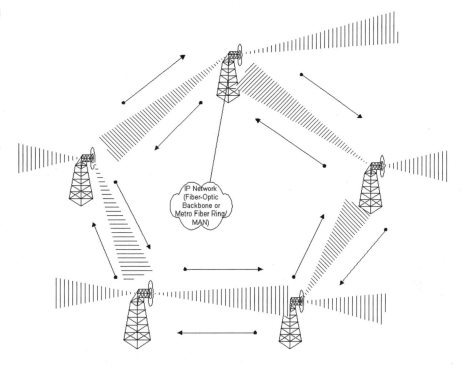

IP Network
(Fiber-Optic
Backbone or
Metro Fiber Ring/
MAN)

addition, users within range of each other form a network, albeit without a connection to a larger network or the Internet.

Peer-to-peer mode is one part of the 802.11 standard. Most WLANs are operating in the infrastructure mode in which multiple users independently connect to APs. This method severely limits the useful range of the network, forcing network administrators to add multiple APs to create an extended coverage area. The software uses the peer-to-peer capabilities included in every 802.11 card to achieve increased network coverage by making all card users a potential part of the transmission network. Figure 3-6 illustrates an ad hoc peer-to-peer network.[9]

[9]Matthew Peretz, "802.11 Coverage for Miles and Miles," 802.11-Planet.com, www.80211-planet.com/news/article/0,,1481_970641,00.html, February 7, 2002.

Figure 3-6
An ad hoc peer-
to-peer network

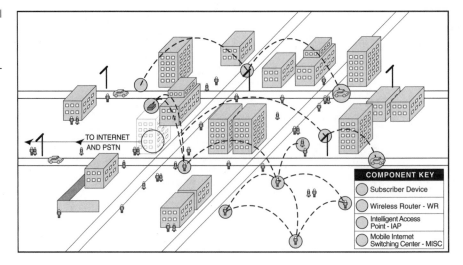

Network Features and Products

Traditional wireless solutions typically attempt to create a mobile broadband network by overlaying some IP equipment onto a circuit-switched, voice-centric system. An ad hoc peer-to-peer network offers an end-to-end, IP-based, packet-switched, mesh architecture that mirrors the wired Internet's architecture and its resulting advantages. In peer-to-peer technology, the users are the network in that they add mobile routers and repeaters (or picocells) to the network infrastructure.

Because users carry much of the network with them, network capacity and coverage is dynamically shifted to accommodate changing user patterns. As people congregate and create pockets of high demand, they also create additional routes for each other, thus enabling access to network capacity through neighboring APs via multihopping. Users will automatically hop away from congested routes and APs to less congested routes and network APs. This permits the network to dynamically and automatically balance capacity and increase network utilization.

The Advantages of Ad Hoc Peer-to-Peer Networks

Ad hoc peer-to-peer networks offer a number of exciting advantages for new market entrants or municipally owned and operated networks. First, the permanent, fixed components such as APs and wireless routers are small and unobtrusive relative to the cell towers found in *third-generation* (3G) architectures. This provides the advantage of much less expensive deployment both in terms of physical plant and legal issues (leasing roof rights, for example). The time needed to deploy service in a given market is also greatly reduced.

Secondly, when enough subscriber devices are present in a given area, the reach of a network is instantly and inexpensively increased. By virtue of using a subscriber device as a router or repeater, the service provider is spared the expense of APs and wireless routers. Furthermore, a network is established among subscriber devices where there is no *intelligent access point* (IAP) or wireless router to connect the subscriber devices to the Internet or other networks.[10]

In ad hoc peer-to-peer mobile architecture, all nodes in the network, including subscriber devices, act as routers and repeaters for other subscribers in the network. This enables users to hop between any number of devices in the network to achieve the desired connection. As a result, user devices can also act as wireless routers. That is, they can act as routers and repeaters for other users. This increases network robustness while reducing infrastructure deployment costs. Ad hoc peer-to-peer networks make it easy for two people to directly share files, e-mail, music, video, or voice calls. Network infrastructure is not needed. Therefore, users can form high-speed voice and data networks anywhere, anytime. Instead of wireless operators subsidizing the cost of user devices (handsets, for example), users actually subsidize and help deploy the network for the operator.

---

[10]"Corporate and Technology Overview," a white paper from MeshNetworks, www.meshnetworks.com/pdf/wp_corpoverview.pdf, 2002.

The Components of Ad Hoc Peer-to-Peer Networks

The network is comprised of the following elements: subscriber devices (including *personal digital assistants* [PDAs], laptops, mobile phones, automobiles, and so on), wireless routers, and APs. Subscriber devices can be either mobile or fixed, whereas the remaining elements are fixed. Wireless routers and APs can be mounted on utility poles, billboards, buildings, or any other convenient structure. It is important to note that the transceiver and modem technology within a subscriber device is identical to the transceiver technology in the fixed infrastructure. This keeps subscriber and infrastructure costs exceptionally low.

Conclusion

The common misperception on 802.11b was that its maximum range was limited to 100 meters. With proper engineering, it can reach 20 miles point to point. The rollout of the 802.16 MAN protocol allows the extension of 802.11b and associated wireless protocols over a wide geographic area. By stepping down from a MAN to lower bandwidth networks, wireless networks can reach out to residential markets and other low-density markets. Ad hoc peer-to-peer networks, by virtue of not requiring expensive infrastructure, are perhaps the most cost-effective means of extending a wireless network. This has the potential to extend the network even further. Here, the subscribers are the network.

Wi-Fi systems act like small routers, with each node relaying to its nearest neighbors. Messages hop peer to peer across a broad interconnected nexus. This produces a broadband telecommunications system, built by separate, independent, interconnecting with each other for their common good. Figure 3-7 graphically illustrates how bandwidth is distributed via a series of interlocking networks.

Two facts make this peer-to-peer structure so interesting. First, its emergence and growth are viral. Viral telecommunications is a truly new, bottom-up phenomenon. This produces a broadband telecom-

Figure 3-7
Extending the
range of
wireless data
transmission via
architecture

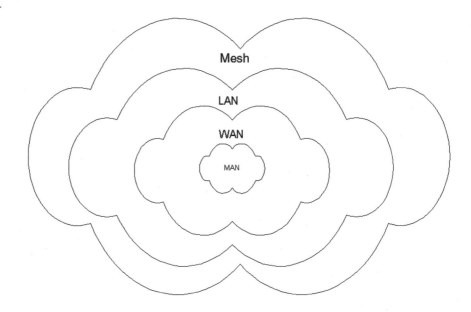

munications system, comprised of separate entities interconnecting
with each other for their common good. Second, its performance increases with the number of nodes. Metcalf's Law states that the
value of a network increases exponentially with the addition of every
new node. In this topology, having more nodes equals better service.
By empowering the subscribers to be the ne twork, the cost to service
providers is drastically reduced. The network could also be commu-
nally owned.[11]

---

[11]Nicholas Negroponte, "Being Wireless," *WIRED Magazine* 10, no. 10 (October 2002):
119.

4

Security and
802.11

Unlike wired systems, which can be physically secured, wireless networks are not confined to the inside of buildings. They can be picked up as far as 1,000 feet outside of the premises with a laptop and a gain antenna. This makes *wireless local area networks* (WLANs) inherently vulnerable to interception.

Knowing this, the 802.11 committee added a first line of defense called *Wireless Equivalency Protocol* (WEP). WEP is an encryption protocol that provides the same level of security that wired cables provide. The standard provides both 40- and 128-bit (really only 104-bit) encryption at the link layer using the RC4 algorithm, which is allowed for export by the U.S. government.

Amazingly, many users do not use this encryption at all. This has led to a new hobby called *war driving* or *LAN jacking* where people pack a wireless-equipped laptop and drive around looking for open networks to surf. However, some people take this same approach to spy on corporations and hack into networks.

Electronics retailer Best Buy Co. ran into trouble in the spring of 2002 when customers who had purchased WLAN cards from Best Buy installed the cards in their laptops before they left the parking lot. The customers noticed unencrypted WLAN traffic that contained customer information and possibly credit card numbers. The Best Buy case is an example of why enterprises should at least encrypt their WLAN traffic with WEP. By the end of 2002, it is expected that 30 percent of enterprises will suffer serious security exposures from deploying WLANs without implementing the proper security.[1]

Recently, several weaknesses have been discovered in the WEP protocol. The 802.11i Task Force is currently working on extensions that will help secure the protocol. According to the Wi-Fi Alliance, formally the *Wireless Ethernet Compatibility Alliance* (WECA), smaller organizations should at least turn on WEP, password protect shared drives and resources, change the network name from the default (the *service set identifier* [SSID]), use *Media Access Control* (MAC) address filtering, use session keys, and use a *virtual private network* (VPN) system. They also suggest that larger organizations should consider additional security methods.

---

[1]Martin Reynolds, "What's Up With Wep?" Strategy, Trends and Tactics, Gartner Group, August 9, 2001.

This chapter seeks to define the risks and known problems with WEP as it has been implemented, and identify what needs to be done to secure a variety of installations based on the level of threat expected for the application.

Basic 802.11 Security and Its Known Problems

When IEEE 802.11b was first defined, its security depended on two basic security mechanisms: the SSID and WEP. Some manufacturers have added MAC address filtering to their products.

Service Set ID (SSID)

The SID is a string used to define a common roaming domain among multiple *access points* (APs). Different SSIDs on APs can enable overlapping wireless networks. The SSID was once thought to be a basic password without which the client could not connect to the network. However, this claim can be easily overridden since APs broadcast the SSIDs multiple times per second and any 802.11 analysis tool such as Airmagnet, NetStumbler, or Wildpackets Airopeek can be used to read it. Because users often configure clients, this so-called password is often widely known.

Should you change your SSID? Absolutely. Although the SSID does not add any layer of security, it should be changed from the default value so that other people do not accidentally use your network.

Wired Equivalent Protocol (WEP)

The IEEE 802.11b standard also defines an authentication and encryption method called WEP to mitigate security concerns. Generally, authentication is utilized to protect against unauthorized access to the network, whereas encryption is used to defeat eavesdroppers who may try to decrypt captured transmissions. 802.11 uses WEP for both encryption and authentication.

Four options are available when using WEP:

- Do not use WEP.
- Use WEP for encryption only.
- Use WEP for authentication only.
- Use WEP for authentication and encryption only.

WEP encryption is based on RC4, which uses a 40-bit key in conjunction with a 24-bit random *initialization vector* (IV) to encrypt wireless data transmissions. (This is why you may see some 802.11b systems labeled as having 64-bit encryption. They are no different than those labeled as having 40-bit encryption keys.) If enabled, the same WEP key must be used on all clients and APs for communication. Most vendors today also offer 128-bit WEP (which uses a 104-bit key). This is a stronger encryption method that makes it more difficult for eavesdroppers to decipher over-the-air transmissions. Although it is not part of the IEEE 802.11b standard, this mode has been implemented on many different vendors' products, some of which are not interoperable.

To prevent unauthorized access, WEP also defined an authentication protocol. Two forms of authentication are defined by 802.11b: open system and shared key. Open system authentication enables any 802.11b client to associate with the AP and skip the authentication process. No authentication of clients or encryption of data occurs. It can be used for public access WLANs, which can be found in coffee shops, airports, hotels, conference centers, and other similar venues where the public is invited to use the network. Typically, the open network authenticates the user using user name password over a secure login web page. For closed networks such as the home or enterprise, this mode can be used when other methods of authentication are provided.

Using shared key authentication, the AP sends a *challenge phrase* to the client radio that is requesting authentication. The client radio encrypts the challenge phrase using the shared key and returns it to the AP. If the AP successfully decrypts it back to the original challenge text, it proves that the client has the correct private key. The client is then allowed to make a network connection.

To the casual observer, it would seem that the shared key authentication process is more secure than the open key authentication process. However, since both the challenge phrase (which was sent in cleartext) and the challenge are available, a hacker can derive the WEP key. Thus, neither open system authentication nor shared key authentication is secure.

Because the 802.11 standard relies on external key management services to distribute the secret keys to each station and does not specify key distribution services, most 802.11 client access cards and APs rely on manual key distribution. This means that the keys remain static unless the network administrator changes them. Obvious problems result from the static nature of the keys and the manual process of key management as changing the keys on each station in a large network can be extremely time consuming. If a station is lost due to theft or accident, the keys will need to be changed on all stations. Furthermore, given the mobility of the population and without a convenient way to manage this task, the network administrator may be under great pressure to accomplish this in a reasonable time frame.

Another concern about the robustness of WEP is that it only provides at most four shared static encryption keys. This means that the four encryption keys are the same for all clients and APs every time a client accesses the network. With enough time, physical proximity, and tools downloaded from the Web, hackers can determine the encryption key being used and decrypt data. Since the whole company is using the same set of keys at any one particular time, it is just a matter of a few hours before enough data is collected to crack a 128-bit key.

Since WEP can be cracked, should you use WEP? If you have nothing else, use WEP to make it more difficult on potential hackers or spammers. You don't want to have your bandwidth stolen for someone else's illegal activities. This is the equivalent of asking "since doors can be picked, should I bother locking the door?"

MAC Address Filtering

Besides the two basic security mechanisms that 802.11 provides, many companies implement MAC address filtering in their products. This mechanism is not flawless either.

The MAC address filter contains the MAC addresses of the wireless *network interface cards* (NICs), which may associate with any given AP. Some vendors provide tools to automate the entry and update processes; otherwise, this is an entirely manual process. A MAC filter is also not very strong security since it is easy to discover known good MAC addresses with a sniffer. Then, using Linux drivers available on the Internet for most 802.11 client access cards, you can configure the sniffed MAC address into the card and gain access to the network. Although not perfectly secure, MAC address filtering is one more layer on the onion—it makes it more difficult for someone to gain access.

The other two steps mentioned by the Wi-Fi Alliance, use of session keys and a VPN system, are good, workable solutions for securing Wi-Fi. In order to understand how much security is needed for a particular application, it is important to understand the threats and potential attacks.

Types of Security Threats

Security Risks

Security can be defined as keeping people from doing things you do not want them to do with, on, or from your data, computers, or peripheral devices. Stored information, the accuracy and value of information, access to internal and external services, and the organization's privacy are at risk. The security risks can come from hackers, criminal intruders, corporate raiders, insiders, contractors, and disgruntled employees. Hackers are typically young hobbyists. "Script Kiddiez" copy well-known attacks from the Internet and run them. More sophisticated hackers understand the underlying protocols and their weaknesses. Criminal intruders may be after access to credit card numbers and checking accounts. Corporate raiders may be after financial information, business plans, and intellectual property. Disgruntled employees, insiders, and contractors are a very serious problem since they are already inside.

WLAN Security Model

Intruders can inflict four major classes of attack on a system: interception, fabrication, modification, and interruption.[2] A fifth class of attacks—repudiation—is an attack against the accountability of information. It is an attack from within the system by either the source entity or the destination entity. Each of these classes of attack can be addressed with a security mechanism. Together, the security mechanisms form a cryptosystem. Table 4-1 describes the five classes of attack.

Table 4-1

Five classes of attack

Attack	On	Solved By
Interception	Confidentiality and privacy	Encryption/decryption
Fabrication	Authenticity	Authentication
Modification Replay Reaction	Integrity	Attacks on Integrity can be solved by digital signatures on every message.
Interruption	Availability	No effective solutions exist for interruption / Denial of Service attacks on availability.
Repudiation	Nonrepudiation	Non-repudication currently still suffers of cases of identity theft.

[2]William Stallings, *Network and Internetwork Security: Principles and Practice* (Englewood Cliffs, NJ: Prentice-Hall International, 1999).

Normal Flow

Under normal circumstances, information is sent from the source to the destination (see Figure 4-1).

Interception

Interception is a passive attack on confidentiality where an intruding entity is able to read the information that is sent from the source entity to the destination entity (see Figure 4-2). Sniffing is an example of an interception attack.

The intruder attempts to learn or make use of information from the system, but does not affect system resources. The identity of the source entity can be intercepted and later used in a masquerade attack, or the intruder may be interested in releasing message contents such as authentication information, passwords, credit card numbers, intellectual property, or other sensitive information. The intruder may also be interested in performing traffic analysis on the

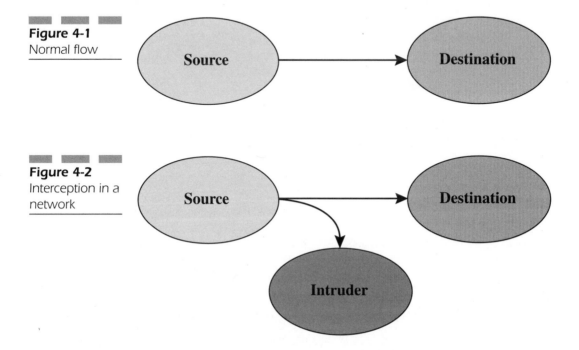

Figure 4-1
Normal flow

Figure 4-2
Interception in a network

system to derive or infer information from the traffic characteristics. The following sections describe examples of interception.

Eavesdropping and Sniffing Eavesdropping is the passive acquisition of information from a network. Just as you can listen to other people's conversations, information can be overheard on the network. This method of gathering information about the network is getting easier with the release of several products. Airopeek, Airsnort, NetStumbler, and WEPCrack are all programs that enable you to acquire information such as the SSID, the MAC address of the AP, and information about whether WEP is enabled.[3]

The nature of a network based on *radio frequency* (RF) leaves it open to packet interception by any radio within range of a transmitter. Interception can occur far outside the user's working range by using high-gain antennas (many of which are standard offerings from some vendors). With readily available tools, an eavesdropper is not limited to just collecting packets for later analysis, but he or she can actually see interactive sessions like web pages viewed by a valid wireless user. An eavesdropper can also catch weak authentication exchanges, such as some web site logins. The eavesdropper could later duplicate the logon and gain access.

The 802.11 standards committee approved WEP, a proprietary encryption design by RSA, before adequate cryptographic analysis was performed. WEP's design has since been analyzed by research teams at Berkeley[4] and the University of Maryland,[5] and serious cryptographic flaws have been found. Researchers at Rice University and AT&T have found an algorithm to crack WEP in about 15 minutes.[6] Hackers have developed tools such as NetStumbler, APSniff, and BSD Airtools to find wireless networks. Tools such as

---

[3]James and Ruth LaRocca, *802.11 Demystified* (New York: McGraw-Hill, 2002), 156–159.

[4]S. Fluhrer, I. Mantin, and A. Shamir, "Weakness in the Key Scheduling Algorithm of RC4," Eighth Annual Workshop on Selected Areas in Cryptography, August 2001.

[5]William A. Arbaugh, Narendar Shankar, and Y. C. Justin Wan, "Your 802.11 Wireless Network Has No Clothes," University of Maryland, www.cs.umd.edu/~waa/wireless.pdf, March 30, 2001.

[6]A. Stubblefield, J. Ioannids, and A. D. Rubin, AT&T Technical Report TD-4ZCPZZ, "Using the Fluher, Martin, and Shamir Attack to Break WEP," Rice University and AT&T Labs, www.cs.rice.edu/~astubble/wep_attack.pdf, August 21, 2001.

WEPCrack[7] and Airsnort[8] can crack WEP regardless of the key length. The 802.11i Task Force is working specifically to correct the flaws in WEP.

WEP is a simple algorithm that uses the RC4 stream cipher to expand a short key and an IV into an infinite *pseudorandom number (PN)* key stream. The sender *exclusive ORs* (XORs) the plaintext (which is appended with a *cyclic redundancy check* [CRC]) with this key stream to produce the ciphertext. The receiver has a copy of this key and uses it to generate an identical key stream. The ciphertext is XORed with the key stream and the original plaintext is recovered. Figure 4-3 illustrates the creation of ciphertext in WEP.

WEP operates at the link layer where packet loss is common. This is why the IV is sent in the clear. If two messages use the same IV and the same key is used with a known plaintext, the other plaintext can be recovered. IEEE 802.11 did not specify how to pick an IV. Most implementations initialize the IV with zero and afterwards increment it by one for each packet sent. This means that if the unit is reset, the IV starts at zero again.

Figure 4-3
The creation of ciphertext in WEP

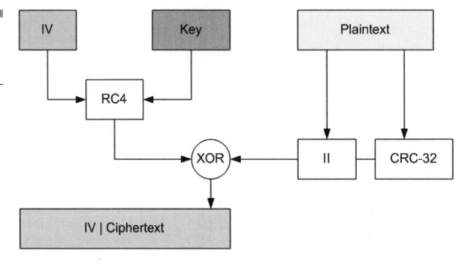

---

[7]WEPCrack, http://sourceforge.net/projects/wepcrack/.

[8]Airsnort, http://airsnort.shmoo.com/.

There are only 24 IV choices. If IVs were randomly chosen, it would only take 12,430 frames to be 99 percent sure that an IV was reused. This is due to the *birthday principle*. For example, in a room of 23 or more people, the probability of two people having the same birthday is 50 percent.

Because WEP sends the IV in the clear along with the encrypted message, it is possible to use dictionary building and statistical methods to crack the WEP key. Both the 64- and 128-bit implementations have the same flaw. The 802.11 standard leaves WEP implementation to the WLAN manufacturers, so the implementations may not be exactly the same. This adds to further weaknesses in the system.

WEP was designed for homes and small businesses. WEP has one static key for the entire system. If a laptop, *personal digital assistant* (PDA), or other 802.11 device is stolen or misplaced from the enterprise, you cannot disable a single user's key; the entire enterprise must be rekeyed.

Another problem is that WEP does not have a key distribution system. In a small business, it is sufficient to enter the keys into the AP and the handful of laptops. However, in a larger organization, manually entering keys is not a scalable operation. If an enterprise needs to be rekeyed, a trusted person must enter the key into the client card of every 802.11 device—manually. The entire enterprise is out of commission until all APs and client cards are updated. Because it is so time consuming to change keys, users tend to use the same key for a long time.

Even if all employees of the enterprise are trusted to administer the key themselves, it still may be difficult for the employees to do. This is because the format of the key varies from one vendor to another. Some vendors use Hex keys, others use ASCII keys, and still others use a key-generation phrase. Some vendors use a combination of two or three of these formats. Some client card vendors have four keys and ask you to choose one out of the four. Asking the users to change to a new key does not work because the stolen laptop will already be preloaded with the keys. To make matters worse, some client cards only hold a single key. The amount of encryption client cards offer is mixed. Some cards do not provide encryption at all (for example, Orinoco Bronze), whereas others only provide 40-bit encryption. Still others allow both 40- and 104-bit encryption.

In many systems, the WEP keys are not properly safeguarded. WEP keys are sometimes stored in the clear. For many APs, WEP keys are sent in the clear from administration terminals into APs using various administration protocols such as the *Simple Network Management Protocol* (SNMP) (version 1 and 2), telnet, and HTTP.

It is clear that a solution that safeguards 802.11 against interception must secure privacy. However, that solution should also solve the associated key distribution problem and properly secure the keys.

Fabrication

Fabrication is an active attack on authentication where an intruder pretends to be the source entity (see Figure 4-4). Spoofed packets and fake e-mails are examples of a fabrication attack.

WEP has two authentication mechanisms. With open system authentication (the default algorithm), the client only announces the intent to associate with the AP, and the AP looks at the *Management Information Base* (MIB) and looks to see if AuthenticationType = OS. If it does, access is allowed. Open system authentication, by its very nature, does not perform authentication and provides no security whatsoever. Figure 4-5 illustrates the open system authentication process.

Figure 4-4
Fabrication in a network

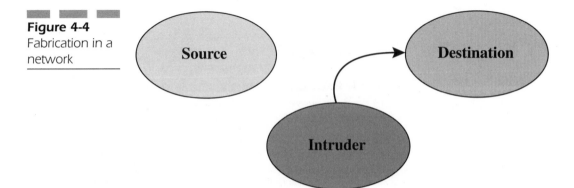

Figure 4-5
Open system
authentication
in an 802.11
network

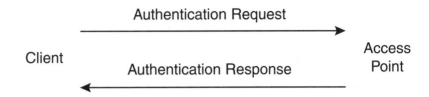

The client connects
to the network.

WEP also has an optional algorithm where the client can ask to be authenticated using shared key authentication. The AP in turn generates a random 128-bit challenge and sends it to the client. The client replies to the challenge, encrypted with the shared secret key, which is configured into both the client and AP. The AP decrypts the challenge, using a CRC to verify its integrity. If the decrypted frame matches the original challenge, the station is considered authentic. Optionally, the challenge/response handshake is repeated in the opposite direction for mutual authentication. Figure 4-6 illustrates the shared key authentication process.

An attacker who captures these frames possesses all of the parts required to derive the RC4 key stream and respond to a future challenge—the plaintext, ciphertext, and IV. The attacker can now pretend he or she is a valid client on the WLAN.

Because the key is shared with all users, no mechanism is available for authenticating individual users and hardware. If the key is leaked out or cracked, anyone who knows the key can use the system. WEP also has no mechanism for the users or hardware to authenticate the AP. Without two-way authentication, it is possible for an attacker to simulate the wireless network and get users to connect to it and reveal additional information that is useful to the attacker.

MAC address filtering is sometimes used to control access to resources. However, MAC address filtering is not adequate for the authentication of users. It is relatively simple to sniff valid MAC addresses out of the air and change the MAC address of a client card to masquerade as a legitimate user. Once access is gained to the network, all computers on the network are accessible because WEP and

Figure 4-6
Shared key
authentication
in an 802.11
network

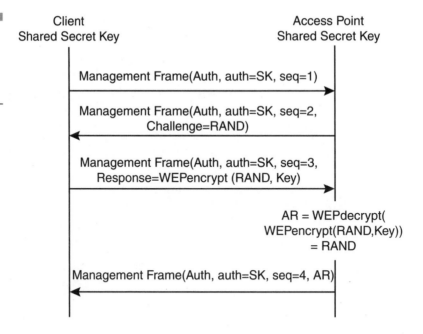

Client
Shared Secret Key

Access Point
Shared Secret Key

Management Frame(Auth, auth=SK, seq=1)

Management Frame(Auth, auth=SK, seq=2,
Challenge=RAND)

Management Frame(Auth, auth=SK, seq=3,
Response=WEPencrypt (RAND, Key)

AR = WEPdecrypt(
WEPencrypt(RAND,Key))
= RAND

Management Frame(Auth, auth=SK, seq=4, AR)

802.11 do not provide access control mechanisms to limit which resources can be accessed. In a home, *small office/home office* (SOHO), or small business environment, this may not be an issue. However, in an enterprise environment, it may be important to control access to resources based on access policies. The following sections provide examples of fabrication.

Man-in-the-Middle Attacks   In order to execute a man-in-the-middle attack, two hosts must be convinced that the computer in the middle is the other host. The classic version of this attack occurs when an attacker intercepts packets from the network, modifies them, and reinserts them back into the network.

Spoofing   Spoofing is the act of pretending to be someone or something that you are not, such as using another person's user ID and password. *Domain Name Service* (DNS) spoofing is accomplished by sending a DNS response to a DNS server on the network. *Internet Protocol* (IP) address spoofing depends on the fact that most routers only look at the destination IP address, not the sending address. Validating the sending IP address can prevent this type of spoofing.

Insertion Attacks   The act of configuring a device to gain access to a network or inserting unauthorized devices into a network in order to gain access is called an *insertion attack*. By installing wireless network cards and being in the vicinity of a target network, a device can be configured to gain access. Unauthorized APs can be installed in an attempt to get users to connect to a hacker's AP rather than to the intended network AP. If these APs are installed behind the corporate firewall, the risk of attack is much greater. This can sometimes be done by well meaning, but misinformed employees.[9]

Brute-Force Password Attacks   Also known as *password cracking* or *dictionary attacks*, this type of attack uses a dictionary and makes repeated attempts to test passwords to gain access to the network. This type of attack is possible even if password authentication is implemented.[10]

Modification

Modification is an active attack on integrity where an intruding entity changes the information that is sent from the source entity to the destination entity (see Figure 4-7). The insertion of a Trojan Horse program or virus is an example of a modification attack.

WEP is wide open to a modification attack without detection because the IV is incremented and CRC is a linear function that only uses addition and multiplication. Thus the following is true:

$$crc(x \oplus y) = crc(x) \oplus crc(y)$$

With the CRC-32 integrity check, it is possible to change one or more bits in the original plaintext and predict which bits in the checksum need to be changed for the message to remain valid. This means it is possible to take messages from the source entity and

---

[9]James and Ruth LaRocca, *802.11 Demystified* (New York: McGraw-Hill, 2002), 157.
[10]Ibid., 157.

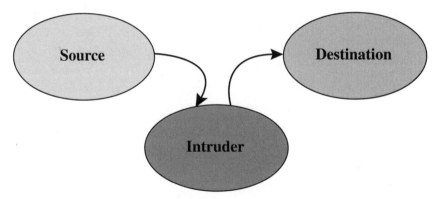

Figure 4-7
Modification
attack in an
802.11 network

modify and reinsert them in the data stream without detection. Basic 802.11 security does not guarantee message integrity. Either WEP or its replacement cipher needs to have a secure integrity check. The following sections provide examples of modification attacks.

Loss of Equipment   The loss of equipment is an issue that has recently received quite a bit of attention due to events within the FBI. The loss of a laptop or other piece of equipment poses the issue of what data is contained within the device. It is possible for an unscrupulous person to dial into the wired network using lost or stolen equipment and stored passwords, and masquerade as an authorized user. This scenario is possible with current wired networks and is not dependent upon having access to a WLAN. The loss of a device equipped with wireless access certainly carries the same risks.

Virus Infection   Virus infection is another issue that affects both wired and wireless networks. To date, there have been no reported viruses that infect cell phones; however, there have been viruses that are capable of sending text messages to cell phones. Two of these are VBS/Timo-A and the LoveBug. There have been reports of viruses that infect Palm OS as well as viruses carried on diskette, CD-ROM, and e-mail. These viruses can infect laptops whether or not they are wireless equipped and can be introduced into and spread via either the larger wired or wireless network.[11]

Replay

Replay is an active attack on integrity where an intruding party resends information that is sent from the source entity to the destination entity (see Figure 4-8).

Basic 802.11 security has no protection against replay. It does not contain sequence numbers or timestamps. Because IVs and keys can be reused, it is possible to replay stored messages with the same IV without detection to insert bogus messages into the system. Individual packets must be authenticated, not just encrypted. Packets must have sequence numbers or timestamps. The following sections describe some examples of replay attacks.

Traffic Redirection   An attacking station can poison the *Address Resolution Protocol* (ARP) tables in switches on the wired network through the AP, causing packets for a wired station to be routed to

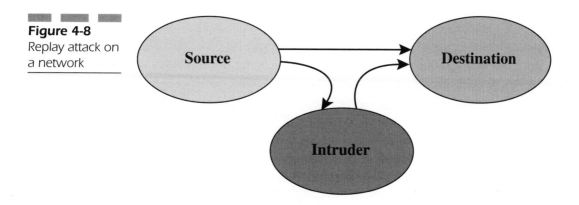

Figure 4-8
Replay attack on a network

---

[11]James and Ruth LaRocca, *802.11 Demystified* (New York: McGraw-Hill, 2002), 153.

the attacking station. The attacker can either passively capture these packets before forwarding them to the attacked wired system or attempt a man-in-the-middle attack. In such an attack, all the susceptible systems could be on the wired network.

Invasion and Resource Stealing Once an attacker has gained the knowledge of how a WLAN controls admittance, he or she may be able to either gain admittance to the network on his or her own, or steal a valid station's access. Stealing a station's access is simple if the attacker can mimic the valid station's MAC address and use its assigned IP address. The attacker waits until the valid system stops using the network and then takes over its position in the network. This enables an attacker to directly access all devices within a network or use the network to gain access to the wider Internet, all the while appearing to be a valid user of the attacked network.[12]

Reaction

Reaction is an active attack where packets are sent by an intruder to the destination (see Figure 4-9). The intruder monitors the reaction. Additional information can be found from this new side channel.

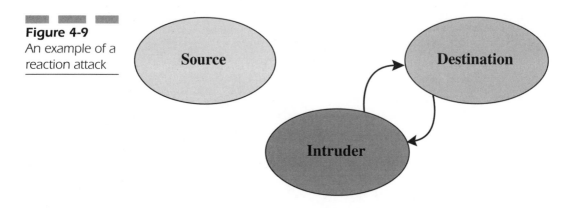

Figure 4-9
An example of a reaction attack

[12]John Vollbrecht, David Rago, and Robert Moskovitz, "Wireless LAN Access Control and Authentication," a white paper from Interlink Networks, www.interlink-networks.com, 2001.

Interruption

Interruption is an active attack on availability where an intruding entity blocks information sent from the originating entity to the destination entity (see Figure 4-10). Examples are *denial of service* (DoS) attacks and network flooding.

The intruder may try to exhaust all network bandwidth using ARP flooding, ping broadcasts, *Transmission Control Protocol* (TCP) SYN flooding, queue flooding, smurfs, synk4, and other flood utilities. The intruder may also use some physical mechanism like RF interference to successfully interrupt a network. A related attack is a degradation of service attack where service is not completely blocked, but the *quality of service* (QoS) is reduced. With basic 802.11 security, little can be done to keep a serious intruder from mounting a DoS attack. The following sections describe some interruption attacks.

Denial of Service (DoS) Attacks   DoS attacks do not allow a hacker to gain access to the network; rather, they basically make computer systems inaccessible by overloading servers or networks with useless traffic so legitimate users can no longer access those resources. The intention is to prevent the network from providing services to anyone. This is usually accomplished by overloading a resource to cause a failure. The overload causes the host to become unavailable, much like those annoying messages stating "all circuits

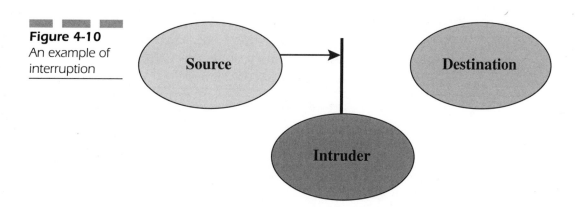

Figure 4-10
An example of interruption

are busy." Many variations of these types of attacks exist depending on the type of resource blocked (disk space, bandwidth, internal memory, and buffers), and some are more easily prevented than others. In the simplest case, turning off the service when it is not needed prevents this type of attack. In other cases, they cannot be easily blocked without limiting the use of a necessary resource. In a wireless network, because the airwaves are shared by other devices such as cordless telephones, microwave ovens, and baby monitors, an attacker with the proper equipment can flood the airwaves with noise and disrupt service to the network.[13]

Rogue Networks and Station Redirection   An 802.11 wireless network is very susceptible to a rogue AP attack. A rogue AP is one owned by an attacker that accepts station connections and then intercepts traffic and might also perform man-in-the-middle attacks before allowing traffic to flow to the proper network. The goal of a rogue is to move valid traffic off the WLAN onto a wired network for attacking (or to conduct the attack directly within the rogue AP) and then reinsert the traffic into the proper network. Such rogue APs could be readily deployed in public areas as well as shared office space areas.

Repudiation

Repudiation is an active attack on nonrepudiation by either the source or the destination where the source entity denies sending a message or the destination entity denies receiving a message (see Figure 4-11).

Basic 802.11 security does not have nonrepudiation. Without nonrepudiation, the source entity can deny ever having sent a message and the destination entity can deny ever having received the message.

---

[13]James and Ruth LaRocca, *802.11 Demystified* (New York: McGraw-Hill, 2002), 153.

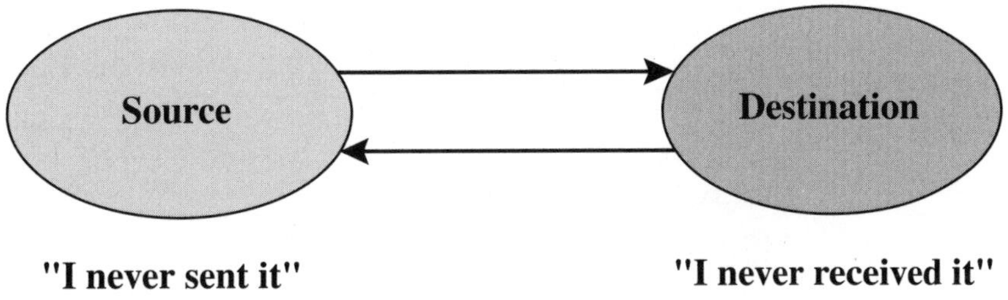

Figure 4-11
An example of
repudiation

Source   Destination

"I never sent it"   "I never received it"

Network Architecture

Typical Network Architecture with WLAN Added

The LAN network should be protected from users on wireless APs. Figure 4-12 shows a typical corporate infrastructure today. The Internet connects to a router on the WAN side. On the LAN side of the router, you may optionally connect a *demilitarized zone* (DMZ) server that is accessible from the Internet for file transfer, for example. A firewall separates the Internet from the corporate network on the LAN side. Often, this firewall function is included in the router.

Figure 4-12
The WLAN
architecture

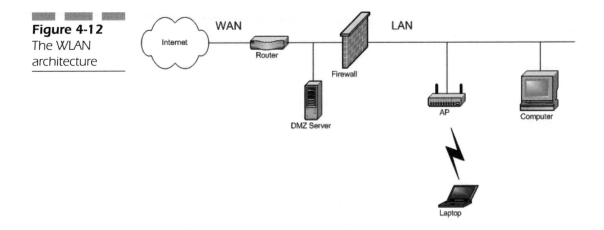

The LAN side serves computers and now more recently APs and wireless laptops. However, the new AP accidentally creates a way to get in behind the firewall through the air link.

Figure 4-12 shows how this leaves the network open to vulnerabilities. Once a user has gained wireless access, he or she also has access to the LAN inside the company.

Typical Network Architecture with a WLAN and Wireless Firewall Added

The network architecture can be changed by adding a wireless authentication firewall that regulates access to the LAN by enabling users to pass only after they have been authenticated, as shown in Figure 4-13. An optional wireless DMZ server or capture portal may exist on the WLAN side of the network. The wireless authentication firewall separates the WLAN from the LAN, thus protecting the enterprise's network from access through the wireless equipment. In an 802.1x/*Extensible Authentication Protocol* (EAP) arrangement, the AP will contain the firewall and an additional *Remote Authentication Dial-In User Service* (RADIUS) server will need to be located on the LAN. In a VPN arrangement, the LAN hosts a VPN server that forms the termination point of the VPN tunnel. Both of the fire-

Figure 4-13

A wireless authentication firewall protects the LAN

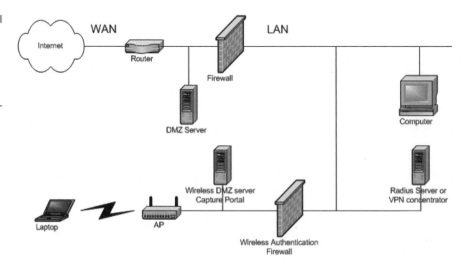

walls will need a hole to carry VPN traffic from the WAN and WLAN side to the LAN.

Mobility and Security

If mobility is used, the solution must be secure during handoff. Handoffs open the network up to a redirection attack. If the network is not properly secured, the intruder can take over the communication with the destination entity after the handoff.

Security Policy—A Range of Options

You need to know what is being protected. These could be devices such as servers, routers, modem banks, and information such as e-mail, intellectual property, trade secrets, customer lists, business plans, and medical records. Sometimes the information has to be protected by law. You also need an idea of who this information is being protected from—hackers, customers, insiders (employees and contractors), and competitors. From this information, a simple risk analysis can be performed to determine what is at risk (data or the network) and the level of countermeasures required to solve the problem. In risk management, you can ignore, accept, defend, or pass on a problem. Unfortunately, there is no canned security policy that you can obtain or use. Each business has its own unique requirements and practices that dictate how implementations are made. Table 4-2 shows the varying levels of security, the configuration, what is secured by the configuration, and what applications such a configuration might be used in.

No Security

No security is like leaving your door wide open. Anyone can come in and use the network access. Basically, it is ignoring potential security problems.

Table 4-2 A range of security options for wireless networks

	Security Level	Configuration	What Is Secured?	Applications
0	No security	Network out of the box and no configuration (no WEP)	Nothing	There is no legitimate unsecured application. Nevertheless, many users operate their equipment in this mode out of the box.
1	Public access	User authentication and must supply VPN through the Internet back to the enterprise	Network access	Hot spots, libraries, coffee shops, hotels, airports, and so on with portability
2	Limited security	40- or 128-bit WEP, MAC *access control list* (ACL), and no broadcast	Some network access and data privacy	Home and SOHO with portability
3	Basic security	*Wi-Fi Protected Access* (WPA) (later 802.11i)	Network access and data privacy	Home, SOHO, and small enterprise with portability
4	Advanced security	802.1x/EAP-X and RADIUS	Network access and data privacy	Enterprise with portability
5	End-to-end security	VPNs such as the *Point-to-Point Tunneling Protocol* (PPTP), PPTPv2, *Layer 2 Tunneling Protocol* (L2TP), Kerberos, and *IP Security* (IPSec)	Network access and data privacy	Special applications, business travelers, telecommuting, business to business, and enterprise with outside users

Public Access

Public access is like handing out keys to everyone you know and trust. NoCat Auth, Sputnik, and Wayport all restrict access to public users by authenticating them. The authentication process protects the network by verifying the access credentials. In some cases, billing is exchanged for granting access to the system. In NoCat Auth's case, access can be removed from people abusing the system. Many of these solutions do not provide a secure tunnel for their users. The data sent over the air is in the clear. Users must provide their own protection against breaches of confidentiality such as using a VPN to tunnel back to their enterprise network.

Basic Access Control

Currently, basic access is like hiding a key under the mat. The key is hidden out there for clever people to find, but the network access and data are accessed to not be worth the trouble to break-in. It's easier for crooks to go where no security is in place.

At minimum, you should turn on WEP, password protect shared drives and resources, change the network name from the default (SSID), use MAC address filtering, and turn off broadcasts if possible. The *Institute of Electrical and Electronics Engineers* (IEEE) is working to remove the key from under the mat with two solutions—WEP enhancements in the short term and WEP replacements in the long term.

802.11 Security Measures Beyond WEP

Wi-Fi Protected Access (WPA)   In November 2002, the Wi-Fi Alliance announced the WPA security standard.[14] It will replace the current—and comparably weak—WEP standard offered now on

---

[14]Wi-Fi Alliance, "Overview—Wi-Fi Protected Access," www.wi-fi.com/OpenSection/pdf/Wi-Fi_Protected_Access_Overview.pdf October 31, 2002.

W-Fi equipment. New equipment shipping with WPA should be available starting in mid-2003. By the fall of 2003, WPA will be a requirement. It is anticipated that many vendors will offer firmware and software upgrades for existing Wi-Fi products that will make them work with WPA. When the products are finally available, they will be marked as being Wi-Fi WPA certified.

WPA uses the *Temporal Key Integrity Protocol* (TKIP), a more hardened encryption scheme than that used in WEP. TKIP uses key hashing (KeyMix) and nonlinear *message integrity check* (MIC). TKIP also uses a rapid-rekeying (ReKey) protocol that changes the encryption key about every 10,000 packets. However, TKIP does not eliminate fundamental flaws in Wi-Fi security. If an attacker hacks TKIP, he or she not only breaks confidentiality, but also access control and authentication.

WPA will work in two different ways, depending on the type of network. In homes and small offices lacking *authentication servers* (ASs), the technology will work in a so-called preshared key mode. Users simply enter the network key to gain access.

In the managed mode, it will work with ASs and will require the support of 802.1x and EAP. 802.1x and EAP enable a client network adapter to negotiate via an AP with a back-end AS using securely encrypted transactions to exchange session keys.

Every device on a wireless network must be upgraded to WPA in order for it to work. If a network is sewn together from several manufacturers' devices and the upgrade for one of the manufacturers is not available yet, you will have to wait to be able to deploy WPA. A mixed network can run with both WPA and WEP installed. However, security in the networks will default to WEP, which offers less protection.

WPA contains many parts of 802.11i. However, some of the key elements are not included such as support for a new encryption algorithm called *Advanced Encryption Standard* (AES), which will replace the RC4-based encryption algorithm when 802.11i becomes available. Migrating to AES encryption will require hardware changes since AES is computationally more complex than RC4. Secure fast handoff preauthentication, secure deassociation and deauthentication, and secure peer-to-peer communications (ad hoc mode) will also follow when 802.11i is released. When 802.11i is a deployed standard, products will be labeled as Wi-Fi WPA2 certified.

802.11 Security Measures Beyond WPA

WPA, when it becomes available, will be a good step toward securing a home or SOHO WLAN. However, for larger enterprises, 802.1x/EAP and VPN are still viable means of securing the WLAN. 802.1x/EAP offers more EAP choices, and a VPN enables workers to access the enterprise's network from any location.

802.1x and EAP—Advanced Security

802.1x provides an authentication framework for WLANs, enabling a user to be authenticated by a central authority. The actual algorithm that is used to determine whether a user is authentic is left open and multiple algorithms are possible. Examples are certificate-based solutions (such as *EAP—Transport Layer Security* [EAP-TLS]), password-based solutions (such as *EAP-One Time Password* [EAP-OTP] and *EAP-Message Digest 5* [EAP-MD5]), smart-card-based solutions (such as *EAP—Subscriber Identification Module* [EAP-SIM]), and hybrids (such as *EAP-Tunneled TLS Authentication Protocol* [EAP-TTLS]) that use both certificates and passwords. Some companies offer their own proprietary EAP solution, such as Cisco's *Lightweight EAP* (LEAP).

802.1x uses EAP, an existing protocol (RFC 2284) that works on Ethernet, Token Ring, or WLANs for message exchange during the authentication process.

802.1x Network Port Authentication   802.1x authentication for WLANs has three main components: the supplicant (usually the client software), the authenticator (usually the AP), and the AS (usually a RADIUS server, although RADIUS is not specifically required by 802.1x).[15] The authenticator connects to the LAN network. Figure 4-14 illustrates this process.

The normal flow for an 802.11x authentication is as follows: The supplicant (in the client) tries to connect to the AP by sending a start

---

[15]IEEE 802.1x, http://standards.ieee.org/getieee802/.

Figure 4-14
802.1x
authentication

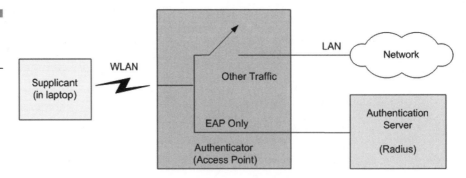

message. The AP detects the supplicant and enables the supplicant's port in an unauthorized state, so only 802.1x/EAP messages are forwarded. All other traffic is blocked.

The supplicant then sends an EAP-start message. The AP relies on an EAP-request identity message to obtain the supplicant's identity. The client's EAP-response packet containing the supplicant's identity is forwarded to the AS.

The AS authenticates the supplicant and either responds by accepting or rejecting the supplicant. Depending on the response from the AS, a positive authentication response will enable the port and negative authentication response will result in the port remaining blocked.

Extensible Authentication Protocol (EAP) To address the shortcomings of WEP for authentication, the industry is working toward solutions based on the 802.1x specification, which is based on the *Internet Engineering Task Force* (IETF) EAP. EAP was designed with flexibility in mind, and it has been used as the basis for many types of network authentication.

802.1x is based on EAP. EAP is formally specified in RFC 2284 and was initially developed for use with the *Point-to-Point Protocol* (PPP). When PPP was first introduced, two protocols were available to authenticate users, each of which required the use of a PPP protocol number. Authentication is not a one-size-fits-all problem, and it was an advanced area of research at the time. Rather than burn up

PPP numbers for authentication protocols that might become obsolete, the IETF standardized EAP.[16]

IEEE 802.1x is not a single authentication method; rather, it utilizes EAP as its authentication framework. This means that 802.1x-enabled switches and APs can support a wide variety of authentication methods, including certificate-based authentication, smart cards, token cards, one-time passwords, and so on. However, the 802.1x specification itself does not specify or mandate any authentication methods. Figure 4-15 illustrates an 802.1x protocol stack showing the use of a variety of EAP types.

Since switches and APs act as a pass through for EAP, new authentication methods can be added without upgrading the switch or AP, by adding software on the host and back-end AS.

Since IEEE 802.1x does not involve encapsulation (unlike *Point-to-Point Protocol over Ethernet* [PPPOE] or VPN), it adds no per-packet overhead and can be implemented on existing switches and APs with no performance impact. This means that IEEE 802.1x can

Figure 4-15
802.1x protocol
stack

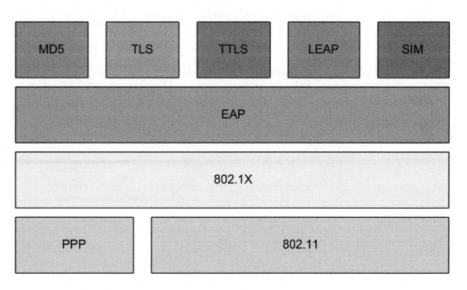

---

[16]Matthew Gast, *802.11b Wireless Networks: The Definitive Guide* (Sebastopol, California: O'Reilly & Associates, 2002), 100.

scale from speeds of 11 Mbps (802.11) to 10+ Gbps and can be enabled on existing switches with a firmware upgrade without the need to buy new hardware. On hosts, since IEEE 802.1x can be implemented in the NIC driver, support can be enabled by obtaining updating drivers from the NIC vendor; you do not need to install a new operating system.

IEEE 802.1x integrates well with open standards for *authentication, authorization,* and *accounting* (AAA) (including RADIUS and *Lightweight Directory Access Protocol* [LDAP]) so it fits in well with the existing infrastructure for managing dial-up networks and VPNs. RADIUS servers (including Windows 2000 IAS) that support EAP can be used to manage IEEE 802.1x-based network access.

These specifications describe how IEEE 802.1x works, and how it can be managed via RADIUS and SNMP. Through RADIUS, IEEE 802.1x permits the management of authorization on a per-user basis. Per-user services include filtering (layer 2 or layer 3), tunneling, dynamic *virtual LANs* (VLANs), rate limits, and so on.[17, 18, 19]

Supplicants for 802.1x/EAP The supplicant for 802.1x/EAP is included in Windows XP. In November 2002, Microsoft not only released a Windows 2000 client,[20] but also pledged future client support for the older Windows 98, ME, and NT 4 releases. It is planned to be developed under Linux, FreeBSD, and OpenBSD with the Open1X project.[21] Apple has not implemented an 802.1x/EAP supplicant at the time of this writing.

802.1x/EAP Authenticators Many commercial APs such as Cisco, Lucent/Orinoco/Agere, and Enterasys feature 802.1x authen-

---

[17]The Unofficial 802.11 Security Web Page, http://www.drizzle.com/~aboba/IEEE/.

[18]Mishra, Arunesh, and Arbaugh, William A, "An Initial Security Analysis of the IEEE 802.1x Standard, Feb. 06, 2002, http://www.cs.umd.edu/~waa/1x.pdf.

[19]Cisco Aironet Response to University of Maryland's Paper, "An Initial Security Analysis of the IEEE 802.1x Standard," March 2002, http://www.cisco.com/warp/public/cc/pd/witc/ao350ap/prodlit/1680_pp.htm.

[20]"Microsoft 802.1x Authentication Client," www.microsoft.com/windows2000/server/evaluation/news/bulletins/8021xclient.asp December 13, 2002.

[21]"Open Source Implementation of IEEE 802.1x," www.open1x.org/.

tication support. Support for homebrew authenticators is provided with the Open1X project.

Remote Access Dial-In User Service (RADIUS) RADIUS[22] is currently the de facto standard for remote authentication. It is a widely deployed protocol for network access AAA in both new and legacy systems. Although a few security and transport issues are associated with it, it is very likely that RADIUS will continue to be widely used for many years to come.[23,24,25] Eventually, RADIUS may be replaced by a new protocol called DIAMETER. RADIUS is simple, efficient, and easy to implement—making it possible for RADIUS to fit into even the most inexpensive embedded devices.

The security issues revolve around the shallowness of the protocol and poor implementation of the specification. The protocol lacks confidentiality, authentication of client messages, and protection against a replay attack (integrity). The shared secret scheme does not allow for enough entropy in the keys and it seems to be open to offline dictionary attacks. RADIUS needs to be secured with an external protocol like IPSec.

The issues with transport are most relevant for accounting, in situations where services are billed according to usage. RADIUS runs on the *User Datagram Protocol* (UDP) and has no defined retransmission or accounting record retention policy, and does not support application-layer acknowledgments or error messages. Lost packets can mean revenue loss. This makes RADIUS accounting unreliable for usage-based billing services, particularly in interdomain usage (such as roaming) where substantial packet loss can occur over the Internet.

Two specifications make up the RADIUS protocol suite: authentication and accounting. The authentication portion can be used to

---

[22]"Remote Authentication Dial-in User Service (RADIUS)," www.ietf.org/rfc/rfc2865.txt, and "Radius Accounting," www.ietf.org/rfc/rfc2866.txt.

[23]Joshua Hill, "An Analysis of the RADIUS Authentication Protocol," www.untruth.org/~josh/security/radius/radius-auth.html.

[24]Bernard Aboba and Ashwin Palakar, "IEEE 802.1x and RADIUS Security."

[25]"Radius Protocol and Best Practices," www.microsoft.com/windows2000/docs/RADIUS_Sec.doc.

determine if a user can gain access to the network. The authentication can be done locally or by proxy to another RADIUS server.

If you're a *wireless Internet service provider* (WISP), and you're reselling access on your APs to roaming parties, you can recognize these roamers by a suffix on their username. The RADIUS server should have a policy implemented as to who, when, and where to proxy to, and the extent to which you allow the remote server's answer to define the response that is sent to the roamer.

The remote server may be chosen based on a part of the username, the network specified in the username, or any other piece of information that's available about a request. You can also get the IP addresses and shared secrets for your remote server from any external source you like, so that you can easily maintain these destinations.

A couple of open-source implementations of RADIUS are available from www.freeradius.org/ and www.xs4all.nl/~evbergen/openradius/index.html. FreeRADIUS is available for a wide range of platforms, including Linux, FreeBSD, OpenBSD, OSF/Unix, and Solaris. OpenRADIUS is a RADIUS server that can be compiled to run on many variations of Unix.

EAP-MD5   EAP-MD5, or the *Challenge Handshake Authentication Protocol* (CHAP),[26] represents a kind of base-level EAP support among 802.1x devices. It is the least secure version of EAP because it uses usernames and passwords for authentication that are easily socialized. It is also vulnerable to dictionary attacks. In addition, EAP-MD5 does not support dynamic WEP keys, which is a critical liability.

LEAP

Cisco was one of the first vendors to market with its proprietary LEAP.[27] LEAP works only with Cisco client 802.11 cards, RADIUS servers, and Cisco APs. LEAP is vulnerable to man-in-the-middle dictionary attacks.

---

[26]"PPP Challenge Handshake Authentication Protocol (CHAP)," www.ietf.org/rfc/rfc1994.txt.

[27]"Authentication with 802.1x and EAP Across Congested WAN Links," www.cisco.com/warp/public/cc/pd/witc/ao350ap/prodlit/authp_an.htm.

EAP-TLS EAP-TLS is an open standard that is supported by many vendors.[28] It uses *Public Key Infrastructure* (PKI) and thus is very secure by using asymmetric public and private keys. EAP-TLS is supported natively in Windows XP and by Windows 2000 severs. The only burden is that a PKI must be set up because every device needs an x.509 certificate. However, once EAP-TLS is set up, it is virtually transparent to the user.

EAP-TTLS EAP-TTLS and EAP-TLS are similar in that both use TLS (the successor to *Secure Sockets Layer* [SSL]) as the underlying strong cryptography. However, EAP-TTLS differs in that only the RADIUS servers, not the users, are required to have certificates. The user is authenticated to the network using ordinary password-based credentials, whose use is made proof against active and passive attacks by enclosing it in the TLS security wrapper.

EAP-SIM EAP-SIM is an EAP method designed by Nokia that allows hardware authentication to a SIM chip. A SIM is a secure processor about the size of a small postage stamp. SIMs are currently used in *Global System for Mobile Communications* (GSM) mobile phones to authenticate a user on a mobile network. After clicking connect and optionally entering a *personal identification number* (PIN), the system authenticates with the network and then connects to the Internet. The beauty of a SIM chip is that it makes cloning authentication secrets very difficult. It is likely that many GSM carriers like T-Mobile and Sonera will implement EAP-SIM to secure their public WLAN offering.

PEAP *Protected EAP* (PEAP) is another Cisco-developed protocol. Although EAP was originally created for use with PPP, it has since been adopted for use with IEEE 802.1x network port authentication. Since its deployment, a number of weaknesses in EAP have become apparent. These include a lack of protection of the user identity or the EAP negotiation, no standardized mechanism for key exchange,

---

[28]"PPP EAP-TLS Authentication Protocol," www.ietf.org/rfc/rfc2716.txt.

no built-in support for fragmentation and reassembly, and a lack of support for fast reconnect.

By wrapping the EAP protocol within TLS, PEAP addresses these deficiencies.[29] Any EAP method running within PEAP is provided with built-in support for key exchange, session resumption and fragmentation, and reassembly. PEAP provides the ability to seamlessly roam between APs. In the near future, PEAP will support the same EAP types that EAP supports.

The Downfall of 802.1x/EAP The downfall of 801.1x/EAP is a lack of supplicants. Although an 802.1x supplicant is supplied in Windows XP, and a patch is available for Windows 2000, no supplicants are available for Windows 98, ME, CE, NT-4, Linux and its variants, and Apple 9 and X operating systems. A Linux open project called open1x has been started at http://www.open1x.org/links.html. Since Apple X is essentially based on Unix, once the open 1x implementation works, it will probably be ported over. The bottom line is that unless supplicants are available that are compatible with the operating systems that are being used, 802.1x/EAP won't be able to provide a complete solution.

VPNs

A VPN enables a specific group of users to access private network data and resources securely over the Internet or other networks. VPNs are characterized by the concurrent use of tunneling, encryption, authentication, and access control over a public network.

VPNs create virtual point-to-point connections using a technique called *tunneling*. As the name suggests, tunneling acts like a pipe that bores through a network cloud to connect two points. Typically started by a remote user, the tunneling process encapsulates data and encrypts it into standard TCP/IP packets, which can then securely travel across the Internet to a VPN server on the other side

---

[29]"Protected EAP Protocol (PEAP)," www.globecom.net/ietf/draft/draft-josefsson-pppext-eap-tls-eap-02.html.

where they are decrypted and de-encapsulated onto the private LAN network.

Two basic VPN types are available:

- **Remote access VPNs**   These securely connect remote users, such as mobile users and telecommuters, to the enterprise. This type of VPN can be used by 802.11 users to initiate a session back to their corporate LAN—for example, salespeople equipped with laptops and telecommuters who would like to connect intermittently from diverse locations such as hotels, airports, convention centers, and coffee shops. The key concerns are encryption and authentication. Performance and bandwidth can also be sacrificed because the connection is broadband.

- **LAN-to-LAN VPNs**   These securely connect remote and branch offices to the enterprise (intranet VPNs). They also securely connect third parties, such as customers, suppliers, and business partners, to the enterprise (extranet VPNs). Intranet or extranet VPNs can be used to secure wireless point-to-point links. For example, a wireless back haul may connect a hot spot back to a central location. This type of VPN needs to be encrypted and authenticated as well as meet strict performance and bandwidth requirements since this kind of connection carries network traffic.

How VPN Works for 802.11   To support 802.11 WLANs, VPN client software application is deployed on all machines that will use the WLAN, and a VPN gateway is introduced into the network between the AP and the WLAN segment, as shown in Figure 4-16.

An encrypted VPN tunnel is built from the laptop through the wireless gateway and terminated at the VPN gateway in order to gain access to the wired LAN through the wireless AP. All traffic passing through the AP must go through the VPN gateway before entering the LAN. The cleartext data on the other side of the secure tunnel can then continue onto its destination inside the local network. The VPN tunnel provides authentication, data confidentiality, and data integrity. Thus, other encryption mechanisms such as WEP are no longer needed.

Figure 4-16
A WLAN with
VPN

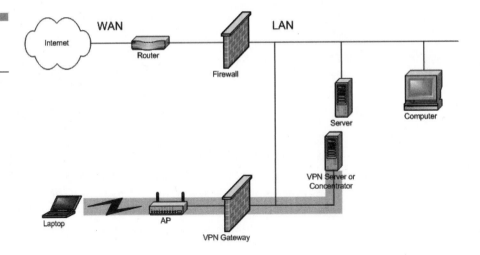

Figure 4-17
A WLAN with
VPN using
remote network
access

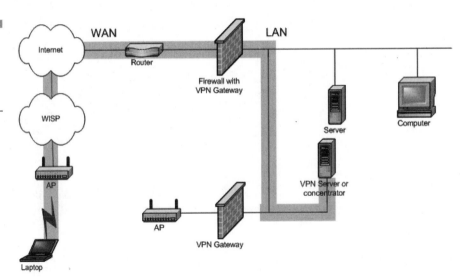

A VPN solution also enables mobile workers to access their network from a remote location where security may or may not be provided, as shown in Figure 4-17.

The secure tunnel extends from the client's computer, through the Internet, and through the firewall/VPN gateway to the VPN server. From the VPN server, the data continues to its destination inside the corporate network.

Vulnerabilities of VPNs   Although VPNs are touted as a secure solution for WLANs, VPNs using one-way authentication are still vulnerable to exploitation such as man-in-the-middle attacks.

The deployment of WLANs in large organizations can create a nightmare of distributing and maintaining client software to all clients. Almost all VPN solutions shipping today are proprietary (not IETF standard) in some form or another and are generally not interoperable. Because of this fact, not all devices may have client software available for any one VPN supplier. Also, it is often the case that once a VPN is installed, a different VPN won't operate on the same machine. Thus, VPNs are impractical for securing a public access WLAN.

It is also important to note that many of the proprietary security extensions may have security flaws due to the lack of cryptographic rigor applied to them. Despite these vulnerabilities, encryption, authentication, and integrity remain essential elements of WLAN security.[30]

VPN Standards   Many protocols have been written for use with VPNs. These protocols attempt to close some of the security holes inherent in VPNs. These protocols continue to compete with each other for acceptance in the industry and are not compatible with each other.

IPSec   IPSec VPNs have nearly become accepted as the de facto standard for securing IP data transmission over shared public data networks since VPN software has been developed for a wide variety of clients. It addresses authentication, data confidentiality, integrity, and key management, in addition to tunneling.

Basically, IPSec encapsulates a packet by wrapping another packet around it. It then encrypts the entire packet. This encrypted stream of traffic forms a secure tunnel across an otherwise unsecured network. Interoperability issues still exist between different vendors' implementations.

---

[30]"Practical Steps to Secure Your Wireless LAN," a white paper from AirDefense, www.airdefense.com.

PPTP PPTP is a protocol specification developed by several companies. Nearly all flavors of Windows include built-in support for the protocol. PPTP was the dominant VPN before IPSec was deployed. PPTP tunnels data, encrypts user data, and authenticates users.

PPTP is a tunneling protocol that provides remote users encrypted, multiprotocol access to a corporate network over the Internet. PPTP uses *generic routing encapsulation* (GRE). PPTP wraps IP packets in GRE packets before sending them down the tunnel. Network layer protocols, such as *Internetwork Packet Exchange* (IPX) and *NetBIOS Enhanced User Interface* (NetBEUI), are encapsulated by the PPTP protocol for transport over the Internet.

The initial releases of PPTP for Windows by Microsoft contained security features that some experts claimed were too weak for serious use. However, Microsoft continues to improve its PPTP support.

L2TP L2TP was developed by Cisco. L2TP supports non-TCP/IP clients and protocols (such as frame relay, *Asynchronous Transfer Mode* [ATM], and *Synchronous Optical Network* [SONET]), but fails to define any encryption standard. Although L2TP is compatible with most network protocols, it is not widely deployed, but is common in certain telco and ISP networks.

L2TP over IPSec L2TP over IPSec offers tunneling, user authentication, mutual computer authentication, encryption, data authentication, and data integrity. L2TP offers multiprotocol support.

SSL SSL, working only with TCP/IP protocols, is the primary protocol for secure connections from web browsers to web servers, usually for secure credit card connections or sensitive data. SSL requires a valid site certificate issued from an authorized certificate authority. SSL provides tunneling, data encryption, mutual authentication, integrity, and nonrepudiation.

SOCKS Network Security Protocol SOCKS is a VPN protocol that operates on layer 5, whereas most others operate at layer 2 or 3. SOCKS version 5 is a circuit-level proxy protocol that was originally designed to facilitate authenticated firewall traversal. Functioning at a higher level means that SOCKS only operates with certain applications. SOCKS v5 supports a broad range of authentication, encryption, tunneling, and key management schemes. It is generally considered to be a market failure.

IP Addresses—Network Address Translation (NAT)/Port Address Translation (PAT) Many VPNs require fully routable/ public IP addresses and no port blocking on any ports except incoming port 80 and port 139 (VPNs do not use these ports). This is because the VPNs cannot tolerate NAT or PAT. However, NAT/PAT is necessary in order to provide network security, prevent subscribers from running host servers on their LAN, and preserve valuable IP address space.

A growing number of VPN clients support emerging standards for UDP encapsulation to push IPSec through NAT/PAT. PPTP often passes through NAT/PAT without trouble, but L2TP over IPSec also requires encapsulation.

Kerberos

Kerberos provides a third method of securing the 802.11 over the air link. It is mainly used by Symbol Technologies, Inc. with their Spectrum24 WLANs. Kerberos provides robust security, uninterrupted network connectivity for voice and data devices, and addresses the security needs and concerns of network managers.

Kerberos provides both user authentication and encryption key management, and can guard networks from attacks on data in transmission, including interruption, interception, modification, and fabrication. Kerberos was voted as the "mandatory-to-implement" security service for 802.11e authentication and encryption key management. Kerberos provides confidentiality, authentication, integrity, access control, and availability. Kerberos also works very well during handoffs between APs resulting in uninterrupted application connectivity. Reauthentication to the network is very quick.

How Kerberos Works for 802.11 Kerberos is based on the key distribution model developed by Needham and Schroeder.[31] Network

---

[31]R. M. Needham and M. D. Schroeder, "Using Encryption for Authentication in Large Networks of Computers," *Communications of the ACM* 21, no. 12 (December 1978): 993–99.

authentication using Kerberos involves four processes: authentication exchange, ticket-granting service exchange, user/server exchange, and secure communications between user and server.[32] This is illustrated in Figure 4-18.

Authentication Exchange The user sends a request to the AS for a ticket to the *ticket-granting server* (TGS). The AS looks up the user in its database, finds the client's secret key, and then generates a session key (SK1) for use between the client and the TGS.

The AS encrypts the session key using the user's secret key to form a message. The AS also uses the TGS's secret key (known only

Figure 4-18
Kerberos
authentication
flow

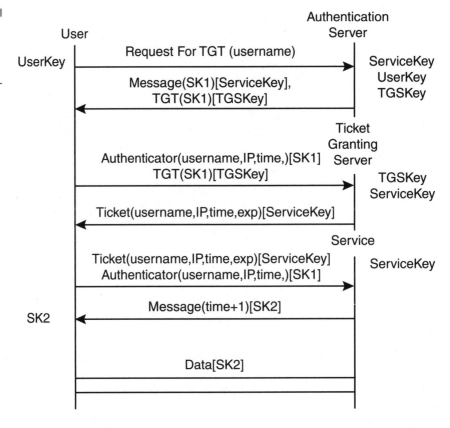

[32]Russel Kay, "Kerberos," Computer World, www.computerworld.com/news/2000/story/0,11280,46517,00.html, July 3, 2000.

to the AS and the TGS) to encrypt the session key and the user's name to form a *ticket-granting ticket* (TGT). The TGT and the message are sent back to the user.

Ticket-Granting Service Exchange   The user decrypts the message and recovers the session key. The user creates an authenticator by encrypting the user's name, IP address, and a timestamp with the session key. The user sends this authenticator, along with the TGT, to the TGS, requesting access to the target server. The TGS decrypts the TGT to recover SK1 and then uses the SK1 inside the TGT to decrypt the authenticator. It verifies information in the authenticator, the ticket, the user's network address, and the timestamp. If everything matches, it lets the request proceed.

Then the TGS creates a new session key (SK2) for the user and target server to use, encrypts it using SK1, and sends it to the user. The TGS also sends a ticket containing the user's name, a network address, a timestamp, and an expiration time for the ticket—all encrypted with the target server's secret key—and the name of the server.

User/Server Exchange   The user decrypts the message and gets SK2. Finally ready to approach the target server, the user creates a new authenticator encrypted with SK2. The user sends the session ticket (already encrypted with the target server's secret key) and the encrypted authenticator. Because the authenticator contains plaintext encrypted with SK2, it proves that the user knows the key. The encrypted timestamp prevents an eavesdropper from recording both the ticket and authenticator and replaying them later. The target server decrypts and checks the ticket, authenticator, user address, and timestamp. For applications that require two-way authentication, the target server returns a message consisting of the timestamp plus one, encrypted with SK2. This proves to the user that the server actually knew its own secret key and thus could decrypt the ticket and the authenticator.

Secure Communications   The target server knows that the user is who he or she claims to be, and the two now share an encryption key for secure communications. Because only the user and target server share this key, they can assume that a recent message encrypted in that key originated with the other party.

The Downfalls of Kerberos   The following are inherent downfalls to the Kerberos authentication system:

- If an attacker is logged on the same computer at the same time as an authorized user, the cached keys located on that computer are accessible to the attacker.

- The Kerberos system relies on the synchronization of the clocks located on the different machines. If an intruder can mislead a host in terms of the correct time, the authentication ticket to the network can be used repeatedly as a result of the nonexpiring timestamp.

- You must trust that all three machines (time/authenticator servers [KDC], the client, and the network server) are void of an intruder.

- If a ticket is forwarded, the system must trust all of the other systems that the ticket has traveled through before reaching the current server. However, the server where the ticket arrives cannot tell where it has come from—it can only tell that it has been on other servers by a flag, which has been set to one.

- Passwords can be guessed by plugging a password guess into the public encryption key algorithm.

- The longer a ticket is granted, the more likely it is to be stolen and used by an unauthorized user.

- In a wireless system using MAC address registration as an authentication method, if the NIC is stolen, the card has the inherent authentication of the user that is tied to that NIC and will be granted access to the network.[33]

Standards   Kerberos V5 is standardized under RFC 1521.[34]

---

[33]S. M. Bellovin and M. Merritt, "Limitations of the Kerberos Authentication System," *Computer Communication Review* 20, no. 5 (1990): 119–132.

[34]The Kerberos Network Authentication Service (V5)," www.ietf.org/rfc/rfc1510.txt.

Conclusion

At the time of this writing, much has been discussed in the press regarding potential security holes in 802.11 security measures. In many instances, 802.11 network managers failed to enable even the most basic security measures built into 802.11. This is the equivalent of leaving a door unlocked and has little to do with the security of 802.11. Any security planning should be an equation figuring in what is to be secured (bank records, military intelligence, jokes from Aunt Nancy, and so on) to what is perceived to be the threat (foreign intelligence services, cyber bank robbers, the casual hacker, an eavesdropping neighbor, and so on) as measured by the resources (financial) available to defend against those perceived threats to network security.

802.11 has a number of built-in measures, including WEP, to protect a network from external threats. Should the network manager feel that WEP is not adequate to protect the network based on the previous equation, a number of other measures can be added to the network to heighten the level of security in the network. This chapter has explored three added security measures in detail. No network is absolutely secure. With the addition of external security measures, 802.11 networks can be as secure as most wired networks.

Interference and Quality of Service (QoS) on 802.11 Networks

If 802.11 networks are to be a viable bypass of the *Public Switched Telephone Network* (PSTN), they must deliver a subscriber experience comparable to that of the PSTN. This is especially important in regards to voice services. Incumbent telcos take great pride in delivering good voice quality on their legacy networks with relatively reliable service. When it comes to replacing the copper wires or fiber cables of the PSTN with the airwaves of 802.11, many people are concerned that the airwaves, since they are not as controllable or predictable as copper wire or fiber cables, will deliver an inferior *quality of service* (QoS) or may be susceptible to interference from other emitters in the electromagnetic spectrum. What will entice consumers, both business and residential, to give up the tried and true PSTN service for wireless service? The answer is greater bandwidth (11 Mbps versus 56 Kbps) that delivers data services including video streaming, video on demand, videoconferencing, music file sharing, and local and long-distance telephone service for a monthly service fee that is marginally more than that of a combined monthly phone, cable TV, and Internet access bill.

Perhaps the main objection to wireless data services is the misperception that the signal will severely suffer from interference from sources that are external to the network. These sources are reputed to be garage door openers, microwave ovens, cordless phones, and so on. A further concern is that the subscriber must have a direct line of sight from the service provider's transmitter. These concerns segue into concerns about the QoS of 802.11 networks. The transmission of a packetized medium (*Internet Protocol* [IP]) over wires has its own series of challenges in addition to interference. This chapter addresses concerns over interference, line of sight, and QoS improvements made possible by 802.11e. There are no problems in telecommunications, only solutions.

As illustrated in Figure 5-1, QoS has to be measured across the total network, encompassing both the wired and wireless portions of the network. It does no good to have a very high-quality *service level agreement* (SLA) for IP services via a wired network if your wireless connection to the *access point* (AP) via a wireless connection suffers from interference or severe latency, and vice versa. Latency is measured from endpoint to endpoint across a network.

Figure 5-1

QoS is measured from endpoint to endpoint, encompassing both the wired and wireless portions of the network.

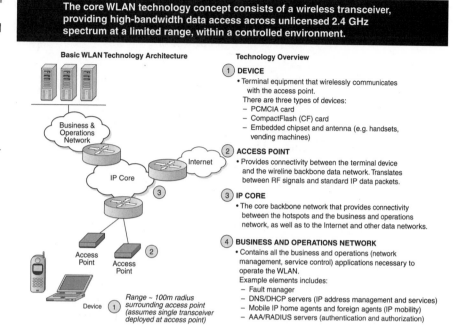

The core WLAN technology concept consists of a wireless transceiver, providing high-bandwidth data access across unlicensed 2.4 GHz spectrum at a limited range, within a controlled environment.

Basic WLAN Technology Architecture

Technology Overview

① DEVICE
• Terminal equipment that wirelessly communicates with the access point.
There are three types of devices:
– PCMCIA card
– CompactFlash (CF) card
– Embedded chipset and antenna (e.g. handsets, vending machines)

② ACCESS POINT
• Provides connectivity between the terminal device and the wireline backbone data network. Translates between RF signals and standard IP data packets.

③ IP CORE
• The core backbone network that provides connectivity between the hotspots and the business and operations network, as well as to the Internet and other data networks.

④ BUSINESS AND OPERATIONS NETWORK
• Contains all the business and operations (network management, service control) applications necessary to operate the WLAN.
Example elements includes:
– Fault manager
– DNS/DHCP servers (IP address management and services)
– Mobile IP home agents and foreign agents (IP mobility)
– AAA/RADIUS servers (authentication and authorization)

Business & Operations Network

Internet

IP Core

Access Point

Access Point

Device

Range ~ 100m radius surrounding access point (assumes single transceiver deployed at access point)

Interference

To most people, QoS in a wireless network refers to interference from other transmission sources. An immediate concern is a profusion of wireless appliances in day-to-day use such as garage door openers, microwave ovens, and cordless phones. The truth is many of these household appliances do not operate on the same frequency as 802.11 or the power of their emission is too low or distant to interfere with 802.11 traffic.

A wide variety of other devices (such as bar code scanners, industrial lighting and industrial heaters, and home microwave ovens) also use the same frequencies. As these *local area networks* (LANs) (and other devices in the *Industrial, Scientific, and Medical* [ISM] band) operate at fairly low power levels, the actual risk of interference is relatively slight, but it does exist. As the popularity of such LANs increases, situations have developed in which such interference has, indeed, become an issue.[1]

External Sources of Interference

Interference can be categorized as having two sources: external and internal. External sources are not related to the 802.11 network and are often categorized as garage door openers, some cordless phones, baby monitors, and so on. Internal sources originate in the 802.11 network. Table 5-1 summarizes the potential external sources of interference in 802.11 networks and their solutions.

Table 5-1

Potential External Sources of Interference to 802.11 Networks and Their Solutions

Source of Interference	Discounting Factor or Solution
Garage door opener	It is on the wrong frequency.
Microwave oven	Commercial microwaves may have the power to generate enough interference to interfere with a WLAN; residential microwaves do not have the power to generate enough interference to be a factor beyond the subscriber's premises.
Cordless phone	Cordless phones are considered to be a nonissue in the industry. They have too little power to interfere beyond the immediate residence or office. If a subscriber's cordless phone is interfering with his or her service, then the subscriber should replace the 2.4 GHz phone with a 900 MHz cordless phone. Why would a residence with a cell phone and *voice over IP* (VoIP) 802.11 service still use a PSTN-connected cordless phone?

---

[1]Ray Horak, "Wireless LANs (WLANs): Focus on 802.11," www.commweb.com/article/COM20020827S0003.

Debunking External Interference Myths Garage door openers are purported to provide interference to 802.11 LANs. This is a myth. Garage door openers operate in the 286 to 390 MHz band; therefore, they don't interfere with 802.11. Many cordless phones operate (900 MHz) in the 802 to 829 MHz ISM band and don't interfere with 802.11 either. However, 2.4 GHz cordless phones do operate on the same band as 802.11 and can cause interference. So how do you deal with interference from other applications of the 2.4 and 5.8 GHz bands since FCC Part 15 users are granted use on a non-interference basis?

The FCC licenses 802.11 wireless APs to operate under Class B, §15.247 of the FCC regulations in the 2.4 GHz ISM band. The regulations state that any device licensed to operate under Part 15 may not interfere with or otherwise disrupt the operation of licensed devices coexisting in the same spectrum. In other words, unlicensed Part 15 devices are the lowest priority. They come after the federal government's FCC-licensed services; Part 18 devices (ISM transmit-only devices) such as telemetry, radiolocation, and *radio frequency* (RF) heating and lighting; and Part 97 Amateur Radio Service. Also, other unlicensed Part 15 devices under the wrong conditions cause interference, such as 2.4 GHz cordless phones, Bluetooth applications, microwave ovens, and 2.4 GHz baby monitors.

Engineering Wireless LANs (WLANs) to Minimize External Interference To minimize external sources of interference, network planners must control the following five parameters:

- The channel/band used
- The distance to the interference (further is better)/the distance to intended signal (closer is better)
- Power levels of interference (lower is better)
- Antenna beam widths
- The protocol used

Changing Channels Sometimes the easiest approach is to change the channel to an unused or lower congested channel. The specifications for both 802.11a and 802.11 stipulate multiple channels or fre-

Table 5-2

The 11
Channels of
802.11

Channel	Frequency (GHz)
1	2.412
2	2.417
3	2.422
4	2.427
5	2.432
6	2.437
7	2.442
8	2.447
9	2.452
10	2.457
11	2.462

quencies. If interference is being encountered on one frequency, then it is merely a matter of switching frequencies to a channel that is not being interfered with. 802.11 provides 11 overlapping channels (for North America), which are 22 MHz wide and centered at 5 MHz intervals (beginning at 2.412 GHz and ending at 2.462 GHz). This means that only 3 channels do *not* overlap (channels 1, 6, and 11). Table 5-2 lists the 11 channels of 802.11 and their frequencies.

802.11a provides 12 channels, which are each 20 MHz wide and centered at 20 MHz intervals (beginning at 5.180 GHz and ending at 5.320 GHz for the lower and middle *Unlicensed National Information Infrastructure* [U-NII] bands, and beginning at 5.745 GHz and ending at 5.805 GHz for the upper U-NII band). It is important to note that none of these channels overlap.[2]

The 5 GHz U-NII band is far less congested, and Wi-Fi has a greater amount of spectrum in which to operate. It also permits more channels. The standards bodies are working on protocols that enable multiple APs to negotiate among themselves automatically for the

---

[2]"A Comparison of 802.11a and 802.11 Wireless LAN Standards," a white paper from Linksys, www.linksys.com/products/images/wp_802.asp.

proper frequency allocation. The 802.11g standard uses *orthogonal frequency division multiplexing* (OFDM) on 2.4 GHz, which is less susceptible to interference and provides more channels. However, the operational range of both 802.11g and 802.11a may be an issue in larger environments.[3]

Once a source of interference has been identified, a common practice among *wireless Internet service providers* (WISPs) is to negotiate and decide which broadcasters (the WISPs) will transmit on what frequency. If such an arrangement cannot be achieved, an WISP can switch to multiple channels in order to avoid interference.

Dealing with Distance   A signal on the same frequency as the 802.11 WLAN, for example, will not cause interference if the source is too distant. That is, the interfering signal becomes too weak to present interference. In addition, if the distance between the AP and the subscriber device is greater than optimal, the signal becomes weak over the distance and becomes susceptible to interference as the interfering signal is greater than the desired signal.

If 802.11 is used as a last-mile solution providing access to a residence or small business, the potential sources of interference must be considered. If the sources of interference (cordless phones or microwave ovens) can be eliminated within the residence or small enterprise, then the second possible source of interference would come from neighboring residences. The potential for those sources of interference are limited by the distance to the subscriber's network and the power level of that interference. Household appliances such as microwave ovens and cordless phones generate too little power to offer interference beyond the building in which they are located unless the device is defective. For example, the door seal may need to be replaced. In this case, the defective microwave oven is a hazard in itself.

Engineering with Power   The power levels of the primary and interfering signals must also be taken into account. If the power level of the interfering signal gets close to the power level of the intended 802.11 or other WLAN signal, then interference will occur. The simplest solu-

---

[3]Chris Fine, "Watch out for Wi-Fi," a report from Goldman Sachs, September 26, 2002, 35.

tion is to increase the power level of the WLAN signal in order to overcome the interfering signal. The limitation here is that the service provider must not interfere with licensed spectrum operators on similar (unlikely) spectrum. The other solution is to reduce the power level of the interfering signal. However, it is important to understand that increasing the power can cause interference for other users of the band and that FCC regulations set legal power output limits.[4]

Antenna Beam Widths   Another way to eliminate interference is to use antennas to shape where the transmitter's signal goes and where the receiver listens. A narrow beam width antenna can increase the effective power toward the receiver and also increase the signal strength of the received signal.

Steerable Antennas   Another engineering approach to overcome QoS issues is to use smart antenna technology. Steering the antenna can do the following for the wireless network: Improve the *signal-to-noise ratio* (SNR) with beam forming, reduce interference due to channel reuse, and mitigate intersymbol interference in multipath environments. Much of this technology falls under the heading of *multiple in, multiple out* (MIMO).

San Francisco-based Vivato is now marketing their Wi-Fi, the first of their kind. Wi-Fi switches deliver the power of network switching with phased-array radio antennas. These Wi-Fi switches use phased-array radio antennas to create highly directed, narrow beams of Wi-Fi transmissions. The Wi-Fi beams are created on a packet-by-packet basis. Vivato calls this technology PacketSteering™.

Unlike current WLAN broadcasting, Vivato's switched beam is focused in a controlled pattern and pointed precisely at the desired client device. These narrow beams of Wi-Fi enable simultaneous Wi-Fi transmissions to many devices in different directions, thus providing parallel operations to many users—the essence of Wi-Fi switching. These narrow beams also reduce co-channel interference, since they are powered only when needed.[5]

---

[4]See FCC Regulations, Part 15.247 and 15.407, www.fcc.gov.

[5]"Vivato Switches Are Changing the Physics of Wireless," a white paper from Vivato, www.vivato.net/prod_tech_technology.html.

Protocol   By using 802.11g instead of 802.11, you gain the advantage of OFDM, which is less susceptible to interference and multipath.

Other Approaches   Other approaches can be taken to control interference. The environment can be controlled so that the interference is limited by controlling the devices used. For example, many airports are requiring all RF applications to be run through their frequency coordinators. The correct band can be chosen for each application. A WISP may consider using 802.11a so that its *wireless wide area network* (WWAN) does not interfere with a possible WLAN solution.

Internal Sources of Interference

Thus far, this chapter has focused on sources of interference that are external to the wireless network. As mentioned in the introduction, a number of challenges arise within a wireless network due to the nature of wireless transmissions. These sources of interference include multipath and channel noise. Both can be engineered out of the network.

Figure 5-2
Signal
interference on
802.11 wireless
networks

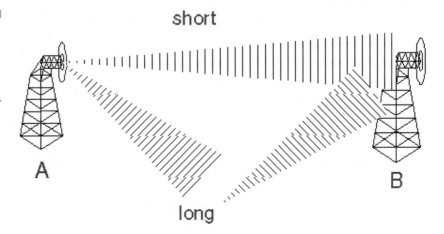

short

long

A

B

Multipath and Fade Margin   Multipath occurs when waves emitted by the transmitter travel along a different path and interfere destructively with waves traveling on a direct line-of-sight path. This is sometimes referred to as *signal fading*. This phenomenon occurs because waves traveling along different paths may be completely out of phase when they reach the antenna, thereby canceling each other. Because signal cancellation is almost never complete, one method of overcoming this problem is to transmit more power. In an indoor environment, multipath is almost always present and tends to be dynamic (constantly varying). Severe fading due to multipath can result in a signal reduction of more than 30 dB. It is therefore essential to provide an adequate link margin to overcome this loss when designing a wireless system. Failure to do so will adversely affect reliability. The amount of extra RF power radiated to overcome this phenomenon is referred to as *fade margin* or *system operating margin*. The exact amount of fade margin required depends on the desired reliability of the link, but 802.11 protocols usually have a 15 to 20 dB fade margin, ensuring a 95 percent confidence interval.[6]

One method of mitigating the effects of multipath is ensuring antenna diversity. Since the cancellation of radio waves is geometry dependent, using two (or more) antennas separated by at least half of a wavelength can drastically mitigate this problem. Upon acquiring a signal, the receiver checks each antenna and simply selects the antenna with the best signal quality. This reduces, but does not eliminate, the required link margin that would otherwise be needed for a system that does not employ diversity. The downside is that this approach requires more antennas and a more complicated receiver design. Another method of dealing with the multipath problem is via the use of an adaptive channel equalizer. Adaptive equalization can be used with or without antenna diversity.

After the signal is received and digitized, it is fed through a series of adaptive delay stages that are summed together via feedback

---

[6]Jim Zyren and Al Petrick, "Tutorial on Basic Link Budget Analysis," a white paper from Intersil, www.intersil.com/data/an/an9804.pdf.

loops. This technique is particularly effective in slowly changing environments such as transmission over telephone lines, but it is more difficult to implement in rapidly changing environments like factory floors, offices, and homes where transmitters and receivers are moving in relation to each other. The main drawback is the impact on system cost and complexity. Adaptive equalizers can be expensive to implement for broadband data links.

Spread spectrum systems are fairly robust in the presence of multipath. *Direct sequence spread spectrum* (DSSS) systems reject reflected signals that are significantly delayed relative to the direct path or strongest signal. This is the same property that enables multiple users to share the same bandwidth in *Code Division Multiple Access* (CDMA) systems. However, 802.11's DSSS does not have enough processing gain and orthogonal spreading codes to do this. *Frequency-hopping spread systems* (FHSS) also exhibit some degree of immunity to multipath. Because an FHSS transmitter is continuously changing frequencies, it always hops to frequencies that experience little or no multipath loss. In a severe fading environment, the throughput of an FHSS system will be reduced, but it is unlikely that the link will be lost completely. OFDM systems such as 802.11a and 802.11g transmit on multiple subcarriers on different frequencies at the same time. Multipath is limited in much the same way in that it is limited in an FHSS system. Also, OFDM specifies a slower symbol rate to reduce the chance a signal will encroach on the following signal, minimizing multipath interference.

Channel Noise   When evaluating a wireless link, three important questions should be answered:

- How much RF power is available?
- How much bandwidth is available?
- What is the required reliability (as defined by the *bit error rate* [BER])?

In general, RF power and bandwidth effectively place an upper bound on the capacity of a communications link. The upper limit in terms of data rate is given by Shannon's Channel Capacity Theorem, as shown in Equation 5-1:

$$C = B \times log2 \, (1 + S/N) \hspace{2cm} (5\text{-}1)$$

where:

C = channel capacity (bits per second)

B = channel bandwidth (hertz)

S = signal strength (watts)

N = noise power (watts)

Note that this equation means that for an ideal system, the BER will approach zero if the data transmission rate is below the channel capacity. In the real world, the degree to which a practical system can approach this limit is dependent on the modulation technique and receiver noise.

For all communications systems, channel noise is intimately tied to bandwidth. All objects that have heat emit RF energy in the form of random (Gaussian) noise. The amount of radiation emitted can be calculated by Equation 5-2:

$$N = kTB \hspace{2cm} (5\text{-}2)$$

where:

N = noise power (watts)

k = Boltzman's constant ($1.38 \times 10\text{--}23$ J/K)

T = system temperature (K), usually assumed to be 290K

B = channel bandwidth (hertz), predetection

This is the lowest possible noise level for a system with a given physical temperature. For most applications, temperature is typically assumed to be room temperature (290K). Equations 5-1 and 5-2 demonstrate that RF power and bandwidth can be traded off to achieve a given performance level (as defined by BER).[7] This implies

---

[7]Ibid.

that using a lower data rate that occupies a lower channel bandwidth will provide better range.

If You Want Interference, Call the Black Ravens

One of the co-author's first real jobs was that of the Intelligence Office, Tactical Electronic Warfare Squadron 135 (abbreviated VAQ-135 with the nickname "World-Famous Black Ravens") of the U.S. Navy. This squadron flew EA-6B tactical jamming aircraft (see Figure 5-3). The airplane is equipped with an ALQ-99 jamming system and is used to jam enemy radar and radio communications in a tactical role. It has been rumored for many years that the squadron's four aircraft, strategically positioned, could shut down most of the electromagnetic spectrum of the United States (TV, radio, and so on).

Figure 5-3
EA-6B tactical jamming aircraft of the U.S. Navy —the "World-Famous Black Ravens." Co-author Franklin Ohrtman is the first person on the left, standing.

In a strategic role during the Cold War, the U.S. Air Force developed the B-52G, which was a bomber equipped with an extensive suite of electronic jamming equipment designed to defeat the Soviet air defenses. This would require overwhelming air defense overlapping radar networks that operated at a variety of frequencies. It would also deliver overwhelming interference on air defense radio communications, making the airwaves unusable for the Soviets. By shutting down Soviet air defense radars and negating their ability to communicate by radio, the B-52G would clear a path for itself and other strategic bombers to targets for destruction by nuclear attack. A trivia question on student exams at the U.S. Navy's Electronic Warfare School in the 1980s asked, "What is the electromagnetic coverage of the B-52G jamming system?" The correct answer was "DC (Direct Current) to daylight."

Given the potential for interference at tactical and strategic military applications relative to the electromagnetic spectrum, it is trivial to negate the benefits of wireless networks based on perceived interference from garage door openers, home microwave ovens, or cordless phones. With proper network engineering, interference from household appliances can be negated.

Line of Sight, Near Line of Sight, and Non Line of Sight

Another objection raised to 802.11 networks is the perception that a subscriber must be located on a direct line of sight from the transmitter. Some detractors fear that this would necessitate an unsightly forest of roof antennas in residential settings not unlike the TV antennas of the 1950s and 1960s. Given the threat to the aesthetics of some residential neighborhoods, some zoning commissions would attempt to outlaw antennas for wireless networks. Others fret that many households would not be able to benefit from wireless broadband as their homes would not be located on a direct line of sight from the service provider's transmitter. Although these are valid considerations, the following paragraphs address the issue of engineering an area to deliver service to the maximum number of subscribers in reach of a service provider's transmitters.

Fresnel Zone and Line-of-Sight Considerations

Line of sight in microwave includes an area around the path called the *Fresnel zone*. The Fresnel zone is an elliptical area immediately surrounding the visual path. It varies depending on the length of the signal path and the frequency of the signal. The Fresnel zone can be calculated and must be taken into account when designing a wireless link. Any object within the Fresnel zone attenuates the transmission path between two points. The maximum radius of the Fresnel zone can be calculated by the following formula:

$$R = 43.3 \times sqrt(d/4f) \hspace{3cm} (5\text{-}3)$$

where d is the distance in miles and f is the frequency in GHz.

For a 5-mile link at 2.4 GHz, the radius is 31.25 feet.

Line of sight means a direct, unobstructed path exists from the transmitter to the receiver. This usually means in any wireless network that the transmission will suffer less degradation than it would if an object(s) was obstructing that path. Line of sight is the best possible configuration for transmission on 802.11 networks.

Non line of sight means the radio link is blocked. However, with proper engineering, it is possible to receive 802.11 services without having a direct line of sight to the service provider's transmitter. This term usually applies where the service provider has deployed its transceivers in a cell network where a backbone services individual cells. If a subscriber is in a location that is non line of sight, he or she will not be served by the WISP. To reach that prospective customer, the service provider would have to deploy a new, costly base station.

One alternative to a new base station would be any-point-to-multipoint technology or an ad hoc peer-to-peer network. In the case of an any-point-to-multipoint network topology, any node already in the network can be used as a relay point to reach the central site (see Figure 5-4). If location 5 is within the line of sight of location 2, for example, node 2 will start functioning as a repeater by simply installing a wide focus antenna connected to its port B. At the new subscriber site (location 5), a transceiver is installed with a directional antenna pointing at location 2.[8]

Ad hoc peer-to-peer technology, also known as *mesh networks*, can also provide a cost-effective means of providing service to non-line-of-sight locations. Single long radio links are replaced with several shorter ones that are less susceptible to noise and multipath. In an ad hoc peer-to-peer network, something as simple as a subscriber device (a handheld *personal digital assistant* [PDA], cell phone lap-

Figure 5-4
Using any-point-to-multipoint technology to reach a non-line-of-sight subscriber

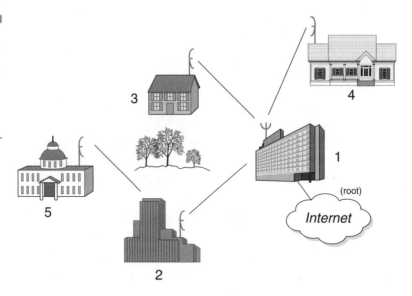

[8]Nuno Bandeira and Larz Poulsen, "Broadband Wireless Network Overcomes Line-of-Sight (LOS) Constraints and Lowers Deployment Cost," a white paper from Wi-LAN, www.wi-lan.com.

top, and so on) can be used as a repeater to reach an AP or base station. The downside is that current client adapters must use new client software to control the routing function of the subscriber device and change it from infrastructure and ad hoc as needed. The cost of APs, base station technology, and wireless routers is becoming less expensive as time goes by. Therefore, a service provider's ability to reach subscribers increases with time. The same goes for potential subscribers if they want to provide the equipment to receive wireless broadband. Not having line of sight to a base station or AP should not prevent them from receiving the benefits of wireless broadband.

The Importance of QoS on 802.11 Networks

When the suggestion is made that 802.11 and the associated protocols could potentially replace the PSTN, one of the first considerations is to provide an alternative to the primary service for which the PSTN was built—voice. A great deal of attention to detail is required to engineer a data network that carries voice. The primary objection to VoIP, the primary means of transmitting voice over a packet network, is that the QoS of an IP network is inadequate to deliver intelligible voice to the subscriber. The limitations of an IP network to deliver adequate QoS for voice and video include latency, jitter, and packet loss. By delivering adequate QoS for voice service, 802.11 presents an alternative to the PSTN's voice services. By delivering good QoS for video delivery, the 802.11 network provides an alternative to a cable or satellite TV service.

In order to best define QoS, this chapter addresses QoS on wired IP networks before describing how many of those concepts apply to wireless networks. IP is the same regardless of whether it is transmitted via a wired or wireless means of transmission. Latency is the chief detractor from QoS in both instances. QoS concerns do not end at the AP. An alternative IP-based network must address QoS end to end.

QoS in Wired IP Networks

A chief objection to VoIP is the notion that the QoS of the VoIP product is inferior to the PSTN. A similar comparison would be made between 802.11 and cable or satellite in the delivery of video services. QoS covers a number of parameters, but is mostly concerned with the users' perception of the voice quality as well as the call setup and keeping the call up for the intended duration of the call (avoiding dropped calls).

Volumes have been written on QoS on IP networks. This chapter sets forth measures that improve QoS on an IP network that makes QoS, especially voice quality over an IP network, as good as that of the PSTN. 802.11 can also offer QoS for video services to compete with cable and satellite services. Earlier in the history of the industry, the perception was that IP voice quality was not as good as that of circuit-switched voice. *International Telecommunication Union—Telecommunication Standardization Sector* (ITU-T) Recommendation E.800 defines QoS as "The collective effect of service performance, which determines the degree of satisfaction of a user of the service." Service providers often state that the QoS end-user experience must equal the PSTN end-user experience. With proper engineering, a VoIP network can deliver QoS that is just as good or better than that of the PSTN.

Factors Affecting QoS in Wired IP Networks

The four most important network parameters for the effective transport of VoIP traffic over an IP network are bandwidth, delay, jitter, echo, and packet loss. Voice and video quality is a highly subjective thing to measure. This presents a challenge for network designers who must first focus on these issues in order to deliver the best QoS possible. This chapter explores the solutions available to service providers that will deliver the best QoS possible. Table 5-3 describes some factors affecting VoIP voice quality.

It is necessary to scrutinize the network for any element that might induce delay, jitter, packet loss, or echo. This includes the hardware elements such as routers and media gateways, and routing pro-

Table 5-3

Factors Affecting VoIP Voice Quality

Factor	Description
Delay	Latency between transmitting an IP packet to receiving the packet at destination.
Jitter	Variation in arrival times between continuous packets transmitted from point A to point B. Caused by packet routing changes, congestion, and processing delays.
Bandwidth	Greater bandwidth delivers better voice quality.
Packet loss	The percentage of packets never received at the destination.

tocols that prioritize voice packets over all other types of traffic on the IP network.

Improving QoS in IP Routers and the Gateway

End-to-end delay is the time required for a signal generated at the caller's mouth to reach the listener's ear. Delay is the impairment that receives the most attention in the media gateway industry. It can be corrected via functions contained in the IP network routers, the VoIP gateway, and engineering in the IP network. The shorter the end-to-end delay, the better the perceived quality and overall user experience. The following sections discuss sources of delay.

Sources of Delay—IP Routers Packet delay is primarily determined by the buffering, queuing and switching, or routing delay of the IP routers. Packet capture delay is the time required to receive the entire packet before processing and forwarding it through the router. This delay is determined by the packet length, link layer operating parameters, and transmission speed. Using short packets over high-speed trunks can easily shorten the delay. VoIP networks use packetization rates to balance connection bandwidth efficiency and packet delay.

Switching or routing delay is the time it takes a network element to forward a packet. New IP switches can significantly speed up the routing process by making routing decisions and forwarding the traffic in hardware devices instead of software. Due to the statistical multiplexing nature of IP networks and the asynchronous nature of packet arrivals, some delay in queuing is required at input and output ports of a packet switch. Overprovisioning router and link capacities can reduce this delay in queuing time.

Sources of Delay—VoIP Gateways   If a voice conversation, for example, has to cross between analog and IP networks, the conversation will have to transit a VoIP gateway. This transition may induce delay in the transmission and degrade the QoS of the conversation. Voice signal processing at the sending and receiving ends, which includes the time required to encode or decode the voice signal from analog or digital form into the voice-coding scheme selected for the call and vice versa, adds to the delay. Compressing the voice signal also increases the delay. The greater the compression, the greater the delay of the packet stream. If bandwidth costs are not a concern, a service provider can utilize G.711, which is uncompressed voice, which imposes a minimum of delay due to the lack of compression.

On the transmit side, packetization delay is another factor that must be entered into the calculations. The packetization delay is the time it takes to fill a packet with data. The larger the packet size, the more time is required. Using shorter packet sizes can shorten this delay, but will increase the overhead because more packets have to be sent, all containing similar information in the header. Balancing voice quality, packetization delay, and bandwidth utilization efficiency is very important to the service provider.[9]

How much delay is too much? Of all the factors discussed in Chapter 4, "Security and 802.11," which outlined the factors that degrade VoIP, latency (or delay) is the greatest. Recent testing by Mier Labs offers a metric for VoIP voice quality. That measure is to determine how much latency is acceptable comparable to the voice quality

---

[9]Bill Douskalis, *IP Telephony: The Integration of Robust VoIP Services*, (Upper Saddle River, NJ: Prentice Hall, 2000), 230–231.

offered by the PSTN. Latency less than 100 *milliseconds* (ms) does not affect toll-quality voice. However, latency over 120 ms is discernable to most callers, and at 150 ms, the voice quality is noticeably impaired, resulting in less than toll-quality communication. The challenge for VoIP service providers and their vendors is to get the latency of any conversation on their network to not exceed 100 ms.[10] Humans are intolerant of speech delays of more than about 200 ms. As mentioned earlier, ITU-T G.114 specifies that delay is not to exceed 150 ms one way or 300 ms round trip. The dilemma is that although elastic applications (e-mail, for example) can tolerate a fair amount of delay, they usually try to consume every bit of network capacity they can. In contrast, voice applications need only small amounts of the network, but that amount has to be available immediately.[11] Figure 5-5 illustrates delay across a network, including delay in a gateway.

Other Gateway Improvements Gateways can be engineered to minimize impairments to QoS. Those impairments are echo, end-to-end delay, buffering delay, and silence suppression. Echo is a phenomenon where a transmitted voice signal is reflected back due to an unavoidable impedance mismatch and a four-wire/two-wire conversion between the telephone handset and the communication net-

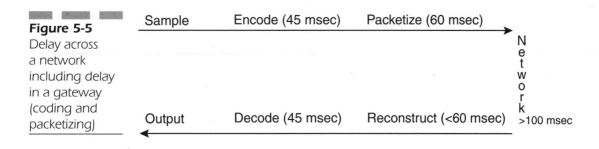

Figure 5-5
Delay across a network including delay in a gateway (coding and packetizing)

---

[10]Mier Communications, "Lab Report — QoS Solutions," www.sitaranetworks.com/solutions/pdfs/mier_report.pdf, February 2001.

[11]John McCullough and Daniel Walker, "Interested in VoIP? How to Proceed," *Business Communications Review* (April 1999): 16–22.

work. Echo can disrupt the normal flow of conversation and its severity depends on the round-trip time delay. If a round-trip time delay is more than 30 ms, the echo becomes significant, making normal conversation difficult. A gateway should have an echo canceller so that when delay reaches above 30 ms, the echo canceller circuits can control the echo.

Occasionally, some cells might arrive late due to a delay in transit. In order to ensure that no under-runs occur, the buffer size should exceed the maximum predicted delay. The size of the buffer translates into delay, as each packet must progress through the buffer at the receiving gateway at the emulated circuit's line rate.

Voice communication is half duplex, which means that one person is silent while the other speaks. A gateway can save bandwidth by halting the transmission of cells at the gateway during these silent periods. This is known as *silence suppression*, or *voice activation detection* (VAD). Gateways can also offer *comfort noise generation*, which approximates the buzz found on the PSTN and lets the user duplicate the PSTN experience with which he or she is most comfortable. Many gateways offer these functions.[12]

Recent research performed by the Institute for Telecommunications Sciences in Boulder, Colorado, compared the voice quality of traffic routed through VoIP gateways with the PSTN. Researchers were fed a variety of voice samples and were asked to determine if the sample originated from the PSTN or the VoIP gateway traffic. The result of the test was that the voice quality of the VoIP gateway-routed traffic was "indistinguishable from the PSTN."[13] It should be noted that the IP network used in this test was a closed network and not the public Internet or other long-distance IP network. This report indicates that quality media gateways can deliver QoS on the same level as the PSTN. The challenge then shifts to ensuring the IP network can deliver similar QoS.

---

[12]"Accelerating the Deployment of Voice over IP (VoIP) and Voice over ATM (VoATM)," a white paper from Telica as posted by International Engineering Consortium (IEC), www.iec.org.

[13]Andrew Craig, "Qualms of Quality Dog Growth of IP Telephony," *Network News* (November 11, 1999): 3.

Improving QoS in the Network

QoS requires the cooperation of all logical layers in the IP network—from application to physical media—and of all network elements from end to end. Clearly, optimizing QoS performance for all traffic types on an IP network presents a daunting challenge. To partially address this challenge, several *Internet Engineering Task Force* (IETF) groups have been working on standardized approaches for IP-based QoS technologies. The IETF's approaches fall into the following categories:

- Prioritization using the *Resource Reservation Protocol* (RSVP) and *differentiated services* (DiffServ)
- Label switching using *Multiprotocol Label Switching* (MPLS)
- Bandwidth management using the subnet bandwidth manager

To greatly simplify the objection that VoIP voice quality is not equal to the PSTN, the problem is overcome by engineering the network to diminish delay and jitter by instituting RSVP, DiffServ, and/or MPLS on the network. The International Softswitch Consortium in its reference architecture recommends RSVP, DiffServ, and MPLS as mechanisms to ensure QoS for VoIP networks.

Resource Reservation (RSVP)   A key focus in this industry is to design IP networks that will prioritize voice packets. One of the earlier initiatives is *integrated services* (IntServ), which was developed by the IETF. It is characterized by the reservation of network resources prior to the transmission of any data. RSVP, which is defined in RFC 2205, is the signaling protocol that is used to reserve bandwidth on a specific transmission path. RSVP is designed to operate with routing protocols such as *Open Shortest Path First* (OSPF) and the *Border Gateway Protocol* (BGP). The IntServ model is comprised of a signaling protocol (RSVP), an admission control routine (determines network resource availability), a classifier (puts packets in specific queues), and a packet scheduler (schedules packets to meet QoS requirements). The latest development is a control/signaling protocol called the *Resource Reservation Protocol—Traffic Engineering* (RSVP-TE). It can be used to establish a traffic-engineered path through the router network for high-priority traffic. This traffic-engineered path can operate independent of other traffic classes.

The IEEE initiative 802.1p is a specification that provides a method to allow preferential queuing and access to media resources by traffic class on the basis of a priority value signaled in the frame. This value provides a consistent method for Ethernet, token ring, or other *Media Access Control* (MAC) layer media types across the sub-network. The priority field is defined as a 3-bit value, resulting in a range of values between 0 and 7, with 0 assigned as the lowest priority and 7 indicating the highest priority. Packets may then be queued based on their relative priority values.[14]

RSVP currently offers two levels of service. The first level is *guaranteed*, which comes as close as possible to circuit emulation. The second level is *controlled load*, which is equivalent to the service that would be provided in a best-effort network under no-load conditions. Table 5-4 covers reservation, allocation and policing.

RSVP works where a sender first issues a PATH message to the far end via a number of routers. The PATH message contains a *traffic specification* (Tspec) that provides details about the data packet size. Each RSVP-enabled router along the way establishes a path state that includes the previous source address of the PATH message. The receiver of the PATH message responds with a *Reservation Request* (RESV) that includes a *flow specification* (flowspec). The flowspec includes a Tspec and information about the type of reservation service requested, such as controlled-load service or guaranteed service.

The RESV message travels back to the sender along the same route that the PATH message took (in reverse). At each router, the requested resources are allocated, assuming that they are available and that the receiver has the authority to make the request. Finally, the RESV message reaches the sender with a confirmation that resources have been reserved.[15]

14Anjali Agarwal, "Quality of Service (QoS) in the New Public Network Architecture," IEEE Canadian Review (Fall 2000): 1.

15Daniel Collins, *Carrier Grade Voice over IP*, (New York: McGraw-Hill, 2002), 362–363.

Table 5-4

Reservation, allocation, and policing are mechanisms available in conventional packet-forwarding systems that can differentiate and appropriately handle isochronous traffic

Reservation, Allocation, and Policing	
RSVP	RSVP provides reservation setup and control to enable the resource reservation that integrated services prescribe. Hoses and routers use RSVP to deliver QoS requests to routers along data stream paths and to maintain router and host state to provide the requested service—usually bandwidth and latency.
***Real-Time Protocol* (RTP)**	RTP offers another way to prioritize voice traffic. Voice packets usually rely on the *User Datagram Protocol* (UDP) with RTP headers. RTP treats a range of UDP ports with strict priority.
***Committed Access Rate* (CAR)**	CAR, a traffic-policing mechanism, allocates bandwidth commitments and limitations to traffic sources and destinations while specifying policies for handling traffic that exceeds the bandwidth allocation. Either the network's ingress or application flows can apply CAR thresholds.

Source: IEEE

Guaranteed service (as opposed to controlled load, see RFC 2212) involves two elements. The first ensures that no packet loss occurs. The second ensures minimal delay. Ensuring against packet loss is a function of the token bucket depth (b) and the token rate (r) specified in the Tspec. At a given router, provided that buffer space of value b is allocated to a given flow and that a bandwidth of r or greater is assigned, there should be little to no loss. Hence, uncompressed voice usually delivers better QoS than compressed voice.

Delay is a function of two components. The first is fixed delay due to the processing within the individual nodes and is only a function of the path taken. The second component of delay is the queuing delay within the various nodes. Queuing is an IP-based QoS mechanism that is available in conventional packet forwarding systems and can differentiate and appropriately handle isochronous traffic. Numerous mechanisms have been designed to make queuing as efficient as possible. Table 5-5 describes these mechanisms.

Table 5-5

Mechanisms
to Increase
Queuing
Efficiency

Queuing	Description
First in, first out (FIFO)	FIFO, also known as the best-effort service class, simply forwards packets in the order of their arrival.
Priority queuing (PQ)	PQ allows prioritization on some defined criteria called a *policy*. Four queues—high, medium, normal, and low—are filled with arriving packets according to the policies defined. *DiffServ code point* (DSCP) packet marking can be used to prioritize such traffic.
Custom queuing (CQ)	CQ permits the allocation of a specific amount of a queue to each class while leaving the rest of the queue to be filled in round-robin fashion. It essentially facilitates the prioritization of multiple classes in queuing.
Weighted fair queuing (WFQ)	WFQ schedules interactive traffic to the front of the queue to reduce response time, and then fairly shares the remaining bandwidth among high-bandwidth flows.
Class-based weighted fair queuing (CBWFQ)	CBWFQ combines CQ and WFQ. This strategy gives higher weight to higher-priority traffic, defined in classes using WFQ processing.
Low-latency queuing (LLQ)	LLQ brings strict PQ to CBWFQ. It gives delay-sensitive data (voice) preferential treatment over other traffic. This mechanism forwards delay-sensitive packets ahead of packets in other queues.

Controlled-load service (see RFC 2211) is a close approximation of the QoS that an application would receive if the data were being transmitted over a network that was lightly loaded. A high percentage of packets will be delivered successfully and the delay experienced by a high percentage of the packets will not exceed the minimum delay experienced by any successfully delivered packet.

Differentiated Service (DiffServ) Another IETF initiative is DiffServ (see RFC 2474). DiffServ sorts packets that require different network services into different classes. Packets are classified at the network ingress node according to SLAs. DiffServ is a set of technologies proposed by the IETF to enable Internet and other IP-based network service providers to offer differentiated levels of service to individual customers and their information streams. On the basis of a DSCP marker in the header of each IP packet, the network routers would apply differentiated *grades of service* (GoSs) to various packet streams, forwarding them according to different *per-hop behaviors* (PHBs). The preferential GoS, which can only be attempted and not guaranteed, includes a lower level of packet latency as those packets advance to the head of a packet queue if the network suffers congestion.[16]

DiffServ makes use of the IP version 4 *Type of Service* (ToS) field and the equivalent IP version 6 Traffic Class field. The portion of the ToS/Traffic Class field that DiffServ uses is known as the DS field. The field is used in specific ways to mark a given stream as requiring a particular type of forwarding. The type of forwarding to be applied is PHB. DiffServ defines two types of PHB: *expedited forwarding* (EF) and *assured forwarding* (AF).

PHB is the treatment that a DiffServ router applies to a packet with a given DSCP value. A router deals with multiple flows from many sources to many destinations. Many of the flows can have packets marked with a DSCP value that indicates a certain PHB. The set of flows from one node to the next that share the same DSCP is known as an *aggregate*. From a DiffServ perspective, a router operates on packets that belong to specific aggregates. When a router is configured to support a given PHB, the configuration is established in accordance with aggregates rather than to specific flows from a specific source to a specific destination.

EF (RFC 2598) is a service in which a given traffic stream is assigned a minimum departure rate from a given node—that is, one that is greater than the arrival rate at the same node. The arrival rate must not exceed a prearranged maximum. This process ensures that queuing delays are removed. As queuing delays are the chief cause of end-to-end delay and jitter, this process ensures that delay and jitter are minimized. The objective is to provide low loss, low delay, and low latency such that the service is similar to a virtual

---

[16]Anjali Agarwal, "Quality of Service (QoS) in the New Public Network Architecture," *IEEE Canadian Review* (Fall 2000): 1.

leased line. EF can provide a service that is equivalent to a virtual leased line.

The EF PHB can be implemented in a network node in a number of ways. Such a mechanism could enable the unlimited preemption of other traffic such that EF traffic always receives access to outgoing bandwidth first. This could lead to unacceptably low performance for non-EF traffic through a token bucket limiter. Another way to implement the EF PHB would be through the use of a weighted round-robin scheduler, where the share of the output bandwidth allocated to EF traffic is equal to a configured rate.

AF (RFC 2597) is a service in which packets from a given source are forwarded with a high probability, assuming the traffic from the source does not exceed a prearranged maximum. If it does exceed that maximum, the source of the traffic runs the risk that the data will be lumped in with normal best-effort IP traffic and will be subject to the same delay and loss possibilities. In a DiffServ network, certain resources will be allocated to certain behavior aggregates, which means that a smaller share is allocated to standard best-effort traffic. Receiving best-effort service in a DiffServ network could be worse than receiving best-effort service in a non-DiffServ network. A given subscriber to a DiffServ network might want the latitude to occasionally exceed the requirements of a given traffic profile without being too harshly penalized. The AF PHB offers this possibility.

The AF PHB enables a provider to offer different levels of forwarding assurances for packets received from a customer. The AF PHB enables packets to be marked with different AF classes and different drop-precedence values within each class. Within a router, resources are allocated according to the different AF classes. If the resources allocated to a given class become congested, then packets must be dropped. Packets that have higher drop-precedence values will be dropped. The objective is to provide a service that ensures that high-priority packets are forwarded with a greater degree of reliability than low-priority packets.

AF defines four classes, which are each allocated a certain amount of resources (buffer space and bandwidth) within a router. Within each class, a given packet can have one of three drop rates. At a given router, if congestion occurs within the resources allocated to a given

AF class, the packets with the highest drop rate will be discarded first so that packets with a lower drop rate value will receive some protection. In order to function properly, the incoming traffic must not have packets with a high percentage of low drop rates. After all, the purpose is to ensure that the highest-priority packets get through in the case of congestion. That cannot happen if all the packets have the highest priority.[17]

In a DiffServ network, the AF implementation must detect and respond to long-term congestion by dropping packets and then respond to short-term congestion, which derives a smoothed long-term congestion level. When the smoothed congestion level is below a particular threshold, no packets should be dropped. If the smoothed congestion level is between a first and second threshold level, then packets with the highest drop-precedence level should be dropped. As the congestion level rises, more of the high drop-precedence packets should be dropped until a second congestion threshold is reached. At that point, all of the high drop-precedence packets are dropped. If the congestion continues to rise, then packets of the medium drop-precedence level should also start being dropped.

The implementation must treat all packets within a given class and precedence level equally. If 50 percent of packets in a given class and precedence value are to be dropped, then that 50 percent should be spread evenly across all packets for that class and precedence. Different AF classes are treated independently and are given independent resources. When packets are dropped, they are dropped for a given class and drop-precedence level. The packets of one class and precedence level might experience a 50 percent drop rate, whereas the packets of a different class with the same precedence level might not be dropped at all. Regardless of the amount of packets that need to be dropped, a DiffServ node must not reorder AF packets within a given AF class, regardless of their precedence level.[18]

---

[17]Collins, 384.

[18]Ibid., 386–387.

MPLS-Enabled IP Networks MPLS has emerged as the preferred technology for providing QoS, traffic engineering, and *virtual private network* (VPN) capabilities on the Internet. MPLS contains forwarding information for IP packets that is separate from the content of the IP header such that a single forwarding paradigm (label swapping) operates in conjunction with multiple routing paradigms. The basic operation of MPLS is to establish *label-switched paths* (LSPs) through the network into which certain types of traffic are directed. MPLS provides the flexibility to form *Forwarding Equivalence Classes* (FECs) and the ability to create a forwarding hierarchy via label stacking. All of these techniques facilitate the operation of QoS, traffic engineering, and VPNs. MPLS is similar to DiffServ in that it marks traffic at the entrance to the network. The function of the marking is to determine the next router in the path from source to destination.

MPLS involves the attachment of a short label to a packet in front of the IP header. This procedure is similar to inserting a new layer between the IP layer and the underlying link layer of the *Open Systems Interconnection* (OSI) model. The label contains all of the information that a router needs to forward a packet. The value of a label can be used to look up the next hop in the path and forward it to the next router. The difference between this routing and standard IP routing is that the match is exact. This enables faster routing decisions in routers.[19]

An MPLS-enabled network, on the other hand, can provide low-latency and guaranteed traffic paths for voice. Using MPLS, voice traffic can be allocated to an FEC that provides the appropriate DiffServ for this traffic type. Significant work has been done recently to extend MPLS as the common control plane for optical networks.[20]

QoS in softswitched networks is corrected with mechanisms similar to those in *time-division multiplexing* (TDM) networks. By engineering out deficiencies in the components (media gateways) and improving the network (DiffServ and MPLS), QoS can be brought up

---

[19]Ibid., 384.

[20]"The Evolution Toward Multiservice IP/MPLS Networks," a white paper from Integral Access, www.integralaccess.com.

to the standards of the PSTN. Although it not as quantifiable as a *Mean Opinion Score* (MOS) on a media gateway, significant progress has been made in recent years in engineering closed IP networks to deliver PSTN-quality voice.

MPLS is not primarily a QoS solution. MPLS is a new switching architecture. Standard IP switching requires every router to analyze the IP header and make a determination of the next hop based on the content of that header. The primary driver in determining the next hop is the destination address in the IP header. A comparison of the destination address with entries in a routing table and the longest match between the destination address and the addresses in the routing table determine the next hop. The approach with MPLS is to attach a label to the packet. The content of the table is specified according to an FEC, which is determined at the point of ingress to the network. The packet and label are passed to the next node, where the label is examined and the FEC is determined. This label is then used as a simple lookup in a table that specifies the next hop and new label to use. The new label is attached and the packet is forwarded.

The major difference between label switching and standard routing based on IP is that the FEC is determined at the point of ingress to the network where information might be available that cannot be indicated in the IP header. The FEC can be chosen based on a combination of the destination address, QoS requirements, the ingress router, or a variety of other criteria. The FEC can indicate information and routing decisions in the network automatically and take that information into account. A given FEC can force a packet to take a particular route through the network without having to cram a list of specific routers into the IP header. This is important for ensuring QoS where the bandwidth that is available on a given path has a direct impact on the perceived quality.[21]

MPLS Architecture   MPLS involves the determination of an FEC value to apply to a packet at the point of the ingress to the network. That FEC value is then mapped to a particular label value and the

---

[21]Collins, 399.

packet is forwarded with the label. At the next router, the label is evaluated and a corresponding FEC is determined. A lookup is then performed to determine the next hop and new label to apply. The new label is attached and the packet is forwarded to the next node. This process indicates that the value of the label can change as the packet moves through the network.

Label-Switching Routers (LSRs) The relationship between the FEC and the label value is a local affair between two adjacent *label-switching routers* (LSRs). If a given router is upstream from the point of view of data flow, then it must have an understanding with the next router downstream as to the binding between a particular label value and FEC.

An LSR's actions depend on the value of the label. The LSR's action is specified by the *Next Hop-Level Forwarding Entry* (NHLFE), which indicates the next hop, the operation to perform on the label stack, and the encoding to be used for the stack on the outgoing link. The operation to perform on the stack might mean that the LSR should replace the label at the top of the stack with a new label. The operation might require the LSR to pop the label stack or replace the top label with a new label and then add one or more additional labels on top of the first label.

The next hop for a given labeled packet might be the same LSR. In such a case, the LSR pops the top-level label of the stack and forwards the packet to itself. At this point, the packet might still have a label to be examined, or it might be a native IP packet without a label (in which case, the packet is forwarded according to standard IP routing).

A given label might possibly map to more than one NHLFE. This might occur where load sharing takes place across multiple paths. Here, the LSR chooses one NHLFE according to internal procedures. If a router knows that it is the next-to-last LSR in a given path, it removes labels and passes the packet to the final LSR without a label. This is done to streamline the work of the last router. If the next-to-last LSR passes a labeled packet to the final LSR, then the final LSR must examine the label, determine that the next hop is itself, pop the stack, and forward the packet to itself. The LSR must then reexamine the packet to determine what to do with it. If the packet arrives without a label, then the final LSR has one less step to execute. The way a particular LSR determines that it is the next-to-last LSR for a given path is a function of label distribution and the distribution protocol used.[22]

Label-Switched Paths (LSPs)   MPLS networks are subsets of a larger IP network. This means points of ingress and egress to the MPLS networks from the larger IP network will exist. An LSR that is a point of ingress to the MPLS network will be responsible for choosing the FEC that should be applied to a given packet. As label distribution works in a downstream-to-upstream direction, an LSR that is a point of egress is responsible for determining a label/FEC binding and passing that information upstream. An LSR will act as an egress LSR with respect to a particular FEC if the FEC refers to the LSR itself, if the next hop for the FEC is outside the label-switching network, or if the next hop involves traversing a boundary. An LSP is the path to a certain FEC from a point of ingress to the egress LSR. The primary function of *label-distribution protocols* (LDPs) is the establishment and maintenance of these LSPs.

Label-Distribution Protocol (LDP)   In the MPLS architecture, the downstream LSR decides on the particular binding. The downstream LSR then communicates the binding to the upstream LSR, which means that an LDP must exit between the two to support such communication. Label distribution is performed in two ways. First, a downstream on demand exists, where a given LSR can request a particular label/FEC binding from a downstream LSR. Second, an unsolicited downstream is available, where a given LSR distributes label/FEC bindings to other upstream LSRs without having been explicitly requested to do so.

MPLS Traffic Engineering   Performance objectives for VoIP networks are either traffic oriented or performance oriented. Traffic-oriented objectives deal with QoS and aim to decrease the impacts of delay, jitter, and packet loss. Performance-oriented objectives seek

---

[22]Ibid.

to make optimum usage of network resources, specifically network bandwidth. Congestion avoidance is a major objective related to both network resource objectives and QoS objectives. In regards to resource objectives, it is imperative to avoid having one part of a network congested while another part of the network is underutilized, where the underutilized part of the network could carry traffic from the congested part of the network. From a QoS perspective, it is necessary to allocate traffic streams where resources are available to ensure those streams do not experience congestion with resultant packet loss and delay.

Congestion occurs two ways. First, the network does not have adequate resources to handle the offered load. Second, the steering of traffic toward resources is already loaded as other resources remain underutilized. The expansion of flow control can correct the first situation. Good traffic engineering can overcome the limitations of steering traffic to avoid congestion. Current IP routing and resource allocation is not well equipped to deal with traffic engineering.

MPLS offers the concept of the traffic trunk, which is a set of flows that share specific attributes. These attributes include the ingress and egress LSRs, the FEC, and other characteristics such as the average rate, peak rate, and priority and policing attributes. A traffic trunk can be routed over a given LSP. The LSP that a traffic trunk would use can be specified. This enables certain traffic to be steered away from the shortest path, which is likely to be congested before other paths. The LSP that a given traffic trunk will use can be changed. This enables the network to adapt to changing load conditions either via administrative intervention or through automated processes within the network.

Traffic engineering on an MPLS network has the main elements of mapping packets to FECs, mapping FECs to traffic trunks, and mapping traffic trunks to the physical network topology through LSPs. The assignment of individual packets to a given FEC and how those FECs are further assigned to traffic trunks are functions specified at the ingress to the network. These decisions can be made according to various criteria, provided they are understood by both the MPLS network provider and the source of packets (customer or other network provider).

A third mapping that must take place revolves around providing the quality that is needed for a given type of traffic. This mapping involves constraint-based routing, where traffic is matched with network resources according to the characteristics of the traffic and characteristics of available resources. That is, one characteristic of traffic is the bandwidth requirement and one characteristic of a path is the maximum bandwidth that it offers.

To date, MPLS is considered the best means of engineering an IP network to handle voice traffic to deliver the best possible QoS. As this technology becomes more widely deployed in IP networks, VoIP will be delivered with a quality that is at least equal to the PSTN.[23]

The Need for QoS in Wireless Networks

An additional network requirement must be supported if the user experience in wireless broadband is to be similar to the user experience in wired broadband (for example, T1 access). The previous paragraphs detailed how wired IP networks can be engineered to limit latency and other factors that detract from QoS. The IEEE has been grappling with the issue of QoS in wireless networks and has recently approved 802.11e, which is backward compatible with other variants of 802.11. That means that improvements in QoS contained in 802.11e can be applied to 802.11 or 802.11a. This section outlines the mechanisms used to ensure the QoS contained in both 802.11 and 802.11e.

Challenges to Wireless QoS

Many previous attempts at WLAN QoS (and non-QoS channel access schemes) show that the strategies that work well in a wired

---

[23]Ibid.

environment do not translate to WLAN. Several factors break possible assumptions: The packet error rate can be in the range of 10 to 20 percent, bit rates vary according to channel conditions, and the "rubber pipe problem" arises where bandwidth managers do not know how much bandwidth they have to manage because a neighboring, unrelated bandwidth manager can take some of it at any time. In addition, if a wireless network is to bypass or substitute for the PSTN, it must be able to prioritize voice and video packets over data packets.[24]

Latency in Wireless Networks

As witnessed earlier in this chapter, the chief threat to an IP network is latency (or delay) of the delivery of packets via the network. Latency is defined as the time it takes for the network to respond to a user command. If latency is high, causing noticeable delays in downloading web pages, then the experience feels nothing at all like broadband, no matter how high the data rates are. Low latency (less than 50 ms) is a requirement for the successful mass-market adoption of wireless services and devices.

The latency experienced by wireless users has a number of contributing sources, including the air link processing, propagation, network processing and transport, the far-end server (if applicable), the application, and the user device. The sum of these latencies must be minimized to ensure a positive end-user experience. Because of the many contributing sources in wired networks, there is little room for latency contributed by the wireless system. Table 5-6 describes the types of delay encountered on an 802.11 network.

The processing delay leads to another very unique disadvantage of systems that are not built on an all-IP basis. Many networks cannot transmit native IP packets and require IP assistance through either a protocol change (transcoding and encapsulation) or the addition of equipment in the network to simulate IP performance. Those mea-

---

[24]A presentation from Intel.

Table 5-6

Types of delay
encountered
on an 802.11
network

Delay	Definition
Air link processing	The time necessary to convert user data to air link packets (code, modulate, and frame user data) and transmit it
Propagation	The time necessary for a signal to travel the distance between the base station and the subscriber device, and vice versa
Network transmission	The time necessary to send the packet across the backhaul and backbone networks, including routing and protocol processing delays and transmission time
Far-end processing	The time required for processing by the far-end servers and other devices

Source: Flarion

sures introduce complexity and packet delays, further impacting the latency of a given system and driving up cost.

Throughput and latency are two essentials for network performance. When taken together, these elements define the speed of a network. Whereas throughput is the quantity of data that can pass from source to destination in a specific time, round-trip latency is the time it takes for a single data transaction to occur (that is, the time between requesting data and receiving it). Latency can also be thought of as the time it takes for data to be sent on one end to be retrieved on the other end (from one user to the other).

Latency is crucial to the broadband experience because the Internet is based on the *Transmission Control Protocol* (TCP). TCP requires the recipient of a packet to acknowledge its receipt. If the sender does not receive a receipt in a certain amount of time (ms), then TCP assumes that the connection is congested and slows down the rate at which it sends packets. TCP is very effective in dealing with congestion on the wired networks.

A system's ability to efficiently handle a large user population depends significantly on its ability to service many small TCP/IP messages per unit time and multiplex many active data users within a given cell. Hence, high latency translates directly into lower system capacity for serving data users, which equates to higher cost. The ideal mobile data network supports both high peak data rates (greater than 3 Mbps) and low packet latency (less than 50 ms). A unique approach is needed to do this over the wireless medium.[25]

QoS in 802.11

The consensus in the industry is that 802.11 alone does not offer adequate QoS. The IEEE has forwarded a new protocol designed to improve QoS in the original 802.11 MAC to enhance support for QoS-sensitive applications such as VoIP, videoconferencing, and streaming video. The original 802.11 MAC included two modes of operation: the *distributed coordination function* (DCF) and the *point coordination function* (PCF). The 802.11e draft specification introduces two new modes of operation: *enhanced DCF* (EDCF) and the *hybrid coordination function* (HCF). As with the original 802.11 MAC, the 802.11e enhancements are designed to work with all possible 802.11 physical layers (the original 802.11, the current 802.11, 802.11a, and 802.11g). The following sections describe QoS efforts in 802.11 and the mechanisms in 802.11e that are designed to improve QoS in wireless networks.[26]

Legacy 802.11 MAC

In order to dissect the progression to 802.11e as a QoS mechanism, it is first necessary to examine the legacy 802.11 MAC. The legacy 802.11 MAC includes support for two access mechanisms: DCF and

---

[25]"Low Latency—The Forgotten Piece of the Mobile Broadband Puzzle," a white paper from Flarion, www.flarion.com.

[26]Priyank Garg, Rushabh Doshi, Majid Malek, Russell Greene, and Maggie Cheng, Stanford University, "Achieving Higher Throughput and QoS in 802.11 Wireless LANs," a white paper from Stanford University, http://milliways.stanford.edu/~radoshi/cs444n/.

PCF. In practice, almost all, if not all, commercial implementations use DCF exclusively. Refer to Chapter 2, "How Does 802.11 Work?" for review of the access mechanisms. Figure 5-6 illustrates the basic access method in DCF and PCF.

Distributed Coordination Function (DCF)

DCF is the basic 802.11 MAC. DCF is based on *carrier sense multiple access with collision avoidance* (CSMA/CA). CSMA/CA is very similar to Ethernet's *carrier sense multiple access with collision detection* (CSMA/CD); however, due to the implementation of the wireless transceiver, collision detection is not possible.

CSMA/CA

The primary QoS mechanism in 802.11 networks is collision avoidance using CSMA/CA. This is the method of sensing the medium and waiting before transmission. In CSMA/CA, stations listen to the medium to determine when it is free. Once a station detects that the medium is free, it begins to decrement its back-off counter (a sort of preemptive back-off). Each station maintains a CW that is used to determine the number of slot times a station has to wait before transmission. The back-off counter only begins to decrement after the medium has been free for a DIFS. If the back-off counter expires and the medium is still free, the station begins to transmit. It is pos-

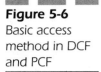

Figure 5-6

Basic access method in DCF and PCF

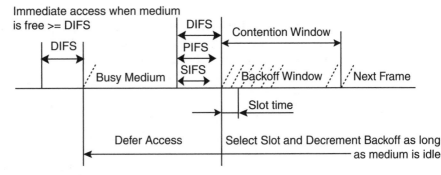

sible that two nodes will begin to transmit at the same time. This causes a collision. Collisions (or other transmission problems) are detected by the lack of an acknowledgement from the receiver. After the detection of a collision, the station randomly picks a new back-off period from its CW (the CW grows in a binary exponential fashion similar to Ethernet) and then attempts to gain control of the medium again. Due to collisions and the binary back-off mechanism, there are no transmit guarantees with DCF. A pictorial representation of the DCF access mechanism is included in Figure 5-6.

Collision Avoidance Mechanisms

In order to avoid collisions, the DCF uses mechanisms for sensing whether the medium is in use before transmitting. If the medium is in use, the station will wait according to a predetermined algorithm before attempting to transmit. The DCF supports complementary physical and virtual carrier sense mechanisms.

Because each medium has different characteristics, physical sensing of the medium is called *clear channel assessment* (CCA). For example, a direct sequence radio PHY can be directed to report the medium to be in use in any of three separate conditions. The first condition reports an in-use condition if any energy above a defined threshold is detected on the medium. The second condition reports an in-use condition if any DSSS signal is detected. The last condition reports an in-use condition if a DSSS signal above a defined threshold is detected on the medium. Physical sensing is very efficient, but it is susceptible to the hidden-node problem (it cannot sense something that is out of range).

In virtual carrier sensing, no actual physical sensing of the medium occurs. Information about the use of the medium is exchanged through the use of control frames. As opposed to physical carrier sensing, virtual sensing greatly reduces the probability of collisions between hidden nodes on a network. It also reduces the overall throughput. This is due to the additional control frames that must be exchanged. Because this overhead is fixed, the smaller the data frames being sent, the higher the percentage of overhead that is added. In networks with a large amount of small packets or low collision rates, it is best to only use physical sensing. For this reason, the DCF's virtual carrier sensing mechanism is optional. The virtual carrier sense control messages are called *Request to Send* (RTS) and *Clear to Send* (CTS) frames. A frame size threshold (RTS threshold) can be set that allows a virtual carrier sense procedure only for packets greater than a specified size. The RTS/CTS procedure is not used for broadcast or multicast frames (single frames with multiple destinations) because this could generate multiple conflicting CTS responses. The virtual carrier sense mechanism also helps to avoid collisions when two overlapping *basic service sets* (BSSs) utilize the same radio channel for transmission.

In Figure 5-7, when node A wants to send data, it sends an RTS frame to the AP with addressing and timing information. It sends the address of the node that will receive the impending data frame

Figure 5-7
The relationship of CTS and RTS in CSMA/CA

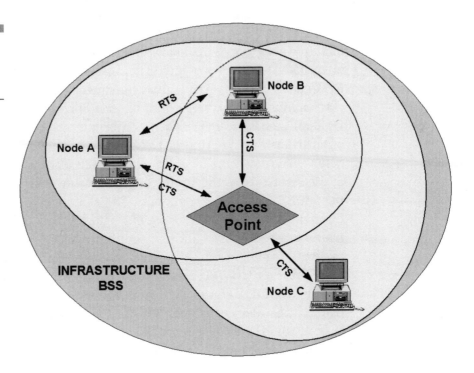

(*receiver address* [RA]), its own address (*transmitter address* [TA]), and how long it wants to transmit (duration). The calculation of the duration has several elements. An AP receiving an RTS frame replies with a CTS frame. The CTS can be heard by all nodes within the AP's range. In forming the CTS frame, the AP copies the TA from the RTS into the RA of the CTS frame. It also copies the duration field into the CTS after adjusting it for the actual transmission of the CTS. The receipt of a CTS causes the receiver to store the duration field as its *network allocation vector* (NAV). The NAV is a timer that indicates the amount of time that remains before the medium can be used. This value counts down on a regular basis, and when it reaches zero, it indicates that the medium is free. It is updated every time an RTS or CTS with a larger value is received. By combining the physical sensing of the medium with the RTS/CTS procedure, it is possible for a hidden node that is unable to receive from the originating node to avoid collisions with an impending data transmission.

In addition to the RTS and CTS control frames, the DCF CSMA/CA procedure requires an *acknowledgement* (ACK) frame to be sent upon successful receipt of certain types of frames. There is no *negative acknowledgement* (NACK), only a timer that indicates how long to wait for an ACK before the transmission is assumed to be in error. The DCF also provides several frame interval timers based on PHY-specific values. These interval timers represent the time that a station must take to sense that the medium is idle before starting a transmission. Two PHY-specific intervals serve as the basis for the other frame interval timers: the slot time and *short interframe space* (SIFS). The slot time for a DSSS PHY (20 ms) is defined as the sum of the receive-transmit turnaround time and the energy-detect time including any propagation delay. The slot time for the IEEE 802.11 frequency-hopping PHY is 50 ms. The SIFS is the shortest of the frame interval spaces and is used to allow the completion of an in-progress transmission. The SIFS for the DSSS PHY is 10 ms. The SIFS for the FHSS PHY is 28 ms.

The slot time and the SIFS are used as components in three other frame intervals. These are the DIFS, the *extended interframe space* (EIFS), and the *PCF interframe space* (PIFS). The DIFS is used by the DCF to enable the transmission of data and management *MAC protocol data units* (MPDUs). The EIFS is used to enable the processing of frames reported to be erroneous by the PHY layer. The PIFS gives a station priority access to the medium when operating in the PCF contention-free mode.

One other timer is used in the DCF virtual CSMA/CA capability: the back-off interval. If a station wanting to transmit detects that a transmission is in progress, it will wait before retrying the transmission. The back-off algorithm determines the time it waits, which is an exponential progression between minimum and maximum values. The starting value for this progression is calculated by multiplying a random number between the minimum and maximum back-off values with the slot time of the PHY. The back-off time is subsequently calculated as sequentially ascending integer powers of 2, minus 1. For example, if the random value is 3 and the slot time is 10 ms, the station would wait 7 or $(2^3-1)\times10$ ms (70 ms). The retries would then continue using 15 (2^4-1) and then 31 (2^5-1) up to the maximum value between retries. Because a random number is used, two stations entering a transmission entry sequence usually do not arrive at the same back-off interval. This prevents two stations from repeatedly colliding because their retry sequences become synchronized. The station also has a retry counter that can limit the number of retries. Figure 5-8 illustrates the virtual carrier sense protocol.

Figure 5-8
Virtual carrier
sense protocol

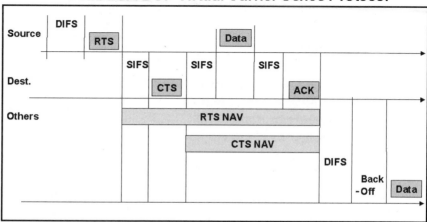

CSMA/CA DCF Virtual Carrier Sense Protocol

The DCF carrier sense protocol is a robust method of overcoming the challenges of radio data transmission between network peers. Centralized traffic management such as that provided by an AP is discussed next.

Data Fragmentation

The longer a transmission lasts, the greater the probability that it will be corrupted by interference. To allow the transmission of shorter frames and reduce the likelihood of interference, the IEEE 802.11 MAC provides a method of breaking transmissions into smaller units. This is called *fragmentation*. A value called the *fragmentation threshold* specifies that frames over a specified size should be divided into multiple transmissions. The frame header contains a sequence control field that shows the order of the fragments. Fragments constituting a frame are transmitted immediately after one another without any contention for the medium. Each fragment has its own *cyclic redundancy code* (CRC), and an individual ACK is transmitted for each fragment. The fragment transmissions are separated by the appropriate frame interval space. The transmission of a sequence of fragments is called a *frame burst*. If an error occurs on a fragment, subsequent fragments are not transmitted until the previous frame is acknowledged. The retransmission and back-off rules apply to fragmented frame transmissions. The NAV is set by duration information in the fragments and ACK. Broadcast and multicast frames are not fragmented even if their size exceeds the fragmentation threshold. Figure 5-9 illustrates frame fragmentation.

Through the use of CSMA/CA and the definition of rules for peer-to-peer as well as centrally managed data transfers, the MAC layer provides reliable structured access to the PHY layer. Physical idiosynchrosies are masked from the upper layers enabling the *Logical Link Control* (LLC) functions and the whole suite of TCP/IP protocols.[27]

---

[27]James and Ruth LaRocca, *802.11 Demystified*, (New York: McGraw-Hill, 2002), 141–142.

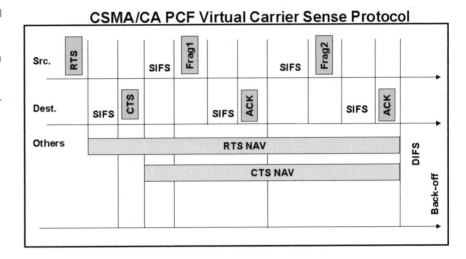

Figure 5-9
Frame fragmentation in virtual carrier sense protocol

Point Coordination Function (PCF)

In an attempt to support limited QoS, 802.11 also defined the PCF. With PCF, the period after each beacon transmission is divided into two sections: the *contention-free period* (CFP) and the *contention period* (CP), which together constitute a superframe. The *point coordinator* (PC) (generally assumed to be co-located at the AP) provides guaranteed access to the medium in the beginning of the CFP by beginning transmission before the expiration of the DIFS. During the CFP, the PC lets stations have priority access to the medium by polling the stations in a round-robin fashion. The CFP is then followed by the CP, during which access to the medium is governed by DCF.

In PCF, the PC has no knowledge of the offered load at each station. The PC simply round-robin polls all stations that have indicated the desire to transmit during the CFP. Any station can request to be added to the poll sequence by a special frame exchange sequence during the CP.[28]

---

[28]Ibid.

IEEE 802.11e—Improved QoS in Wireless Networks

IEEE has formed the 802.11e group to develop improvements to the original 802.11 MAC to enhance support for QoS-sensitive applications such as VoIP, videoconferencing, and streaming video. The original 802.11 MAC included two modes of operation: DCF and PCF.

The 802.11e draft specification introduces two new modes of operation: EDCF and HCF. As with the original 802.11 MAC, the 802.11e enhancements are designed to work with all possible 802.11 physical layers (the original 802.11, the current 802.11, 802.11a, and 802.11g).[29]

EDCF defines eight traffic classes. Various parameters governing back-off can be individually set per traffic class. Medium access is similar to DCF with the addition of *arbitration interframe space* (AIFS). The station cannot begin decrementing the back-off timer until after AIFS. Within a node, each traffic class has a dedicated queue. Traffic class queues contend for access to the virtual channel. Frames that gain access to the virtual channel then contend for medium.

HCF is analogous to PCF, but it allows a *hybrid coordinator* (HC) to maintain state for nodes and allocate contention-free transmit opportunities intelligently. The HC uses the offered load per traffic class at each station for scheduling.

802.11e MAC Enhancements

Many priority schemes to support QoS are currently being discussed. IEEE 802.11 Task Group E currently defines enhancements to the previously discussed 802.11 MAC that are called 802.11e. These enhancements introduce two new MAC modes—EDCF and HCF. Both these QoS-enhanced MAC protocols support up to eight priority levels of traffic that map directly to the RSVP protocol and other protocol priority levels.

[29]Ibid.

Enhanced Distributed Coordination Function (EDCF)   The major enhancement provided by EDCF versus DCF is the introduction of eight distinct traffic classes. Aside from this, EDCF, as the name suggests, works in a fashion similar to the DCF MAC, except that some of the elements of the MAC are parameterized on a per-class basis. Figure 5-10 illustrates the functioning of EDCF.

Here, each traffic class starts a back-off after detecting that the channel is idle for an AIFS. The AIFS is at least as large as the DIFS and can be chosen individually for each traffic class. This is the first per-class MAC parameter added in EDCF.

Second, the minimum value of the CW for each traffic class, denoted by CWMin can be selected on a per-traffic-class basis. In DCF, a global constant CWMin is used to initialize all CW values.

Third, when a collision is detected and the CW has to be increased, the value of CW is increased by a persistence factor, which is also determined on a per-traffic-class basis. A value of 1 for the persistence factor gives a CW that stays constant even in the case of collisions, whereas a value of 2 (which is the default) gives a binary exponential back-off identical to DCF. To calculate the CW in case of a collision, use Equation 5-4:

$$newCW[TC] > = ((oldCW[TC]) + 1) \times PF[TC]) - 1 \quad (5\text{-}4)$$

The CWMax value sets the maximum possible value for the CW on a per-traffic-class basis; however, CWMax is typically intended to remain the same for all traffic classes (at the default valued used in DCF).

Figure 5-10
EDCF and HCF
in 802.11
networks

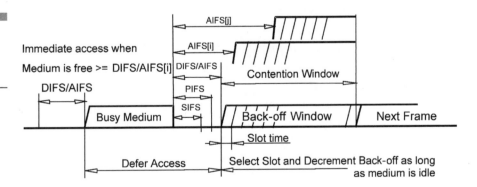

Within a station, the eight traffic classes have independent transmission queues. These behave as virtual stations with the previously mentioned parameters determining their ability to transmit. If the back-off counter of two or more parallel traffic classes in a single station reaches zero at the same time, a scheduler inside the station treats the event as a virtual collision. The *transmit opportunity* (TXOP) is given to the traffic class with the highest priority of the colliding traffic classes, and the others back off as if a collision on the medium occurred.

The QoS parameters, which are provided on a per-traffic-class basis, can be adapted over time. The base station does this by announcing them periodically via the beacon frames, which are transmitted at the beginning of every superframe.

Hybrid Coordination Function (HCF)   HCF is an extension of the polling idea in PCF. Just like in PCF, under HCF, the superframe is divided into the CFP that starts with every beacon and the CP. During the CP, access is governed by EDCF, though the HC (generally co-located at the AP) can initiate HCF access at any time (due to its higher priority, it can begin transmitting before the expiration of the DIFS).

During the CFP, the HC issues a QoS CF-Poll to a particular station to give it a TXOP. The HC specifies the starting time and maximum duration as part of the CF-Poll frame. During the CFP, no stations attempt to gain access to the medium, so when a CF-Poll is received, they assume a TXOP and transmit any data they have. The CFP ends after the time announced by the beacon frame or by a CF-End frame.

If a station is given a CF-Poll, it is expected to start responding with data within an SIFS period. If it does not, the HC can take over the medium after a PIFS time and allocate another CF-Poll to another station. This allows very efficient use of the medium during the CFP.

To determine which station to give the TXOP to, the HC uses per-station/per-traffic-class queue length data that it collects and maintains to reflect the current snapshot of the infrastructure BSS. The QoS Control field that has been added to the MAC frame definition enables stations implementing 802.11e to send queue lengths per traffic class to the HC.

Scheduling The MAC defines protocols and mechanisms to perform HCF and EDCF. However, a couple of opportunities exist to perform scheduling decisions that are not determined by pure random number selection as in DCF.

HC Scheduling The HC has a snapshot view of the per-traffic-class/per-station queue length information over time, including that of the AP itself. With this, it has to decide to whom to allocate TXOPs during the CFP. This involves considering at least the following:

- The priority of the traffic class
- The required QoS for the traffic class (low jitter, high bandwidth, low latency, and so on)
- Queue lengths per traffic class
- Queue lengths per station
- The duration of TXOP available and to be allocated
- Past QoS seen by the traffic class

The practice is to implement a simple scheme of calculating a weighted average queue length per station (weights are based on traffic class queues within a station) and allocate the maximum available TXOP within the CFP to the station with the largest average. However, various possible schemes are available to meet different goals.

Endpoint Scheduling Within TXOP When a wireless station gets a TXOP by polling from the HC, the HC does not specify a particular traffic class for the TXOP. This leaves the decision of which traffic class to service in the TXOP up to the wireless station. This decision can depend on the same factors as for the HC scheduler, except the multistation cell-wise aggregation that the HC scheduler uses is not applicable.

By decentralizing this decision, the protocol provides a scalable mechanism of maintaining traffic class history and servicing as per the QoS seen in the past without collecting this information and detail at the AP, making it unwieldy. A simple scheme that always sends data for the highest-priority traffic class pending during the TXOP has currently been implemented.

EDCF and HCF: QoS in 802.11 Networks

EDCF provides significant improvements for high-priority QoS traffic; however, these improvements are typically provided at the cost of worse performance for lower-priority traffic. It also appears that the EDCF parameters can require significant tuning to achieve performance goals. Despite these problems, EDCF is attractive compared to HCF because of its simplicity.

HCF, just like its predecessor PCF, provides for much more efficient use of the medium when the medium is heavily loaded. Unlike PCF, HCF does a good job of channel utilization even when the channel is operating well below capacity. Due to reduced overhead, HCF can provide better QoS support for high-priority streams while allocating reasonable bandwidth to lower-priority streams.

Both coordination functions are backward compatible with DCF and PCF. This fact, along with our results, leads us to believe that EDCF and HCF will soon see ubiquitous adaptation into mainstream WLAN technology.[30]

Conclusion

802.11e is based on over a decade of experience in designing WLAN protocols and was built from the ground up for real-world wireless conditions. 802.11e is backward compatible with 802.11—

---

[30]Priyank Garg, Rushabh Doshi, Majid Malek, Russell Greene, and Maggie Cheng, "Achieving Higher Throughput and QoS in 802.11 Wireless LANs," a white paper from Stanford University, http://milliways.stanford.edu/~radoshi/cs444n/.

that is, non-802.11e terminals can receive QoS-enabled application streams.

This chapter described measures aimed at improving QoS in 802.11 networks. These wireless networks are potentially capable of delivering QoS comparable to the PSTN. It should be noted that the *Regional Bell Operating Companies* (RBOCs) have been losing phone lines to cell phone service providers at an alarming rate (for the RBOCs) over the last year. In fact, the RBOCs have recorded, percentage-wise, their first decline in lines in use since the Great Depression. Cell phone service is admittedly inferior in quality to that of the PSTN. The motivating factor for landline customers to drop their service from the RBOC is the convenience offered by the cell phone as well as certain price advantages (free long distance in off-peak hours). The point here is that, ultimately, the QoS of the PSTN is not an absolute requirement for consumers. Consumers, in the case of cell phones, have traded off QoS for convenience and price. The PSTN is doomed if it must compete with 802.11 in that 802.11, using 802.11e, potentially delivers a comparable QoS in both voice and data services while offering data rates up to 11 Mbps (compared with most *Digital Subscriber Line* [DSL] plans at 256 Kbps). Given that consumers will trade QoS for convenience and price, as witnessed by the loss of lines to cell phone service providers, it is not hard to imagine they would, trade off the PSTN for the convenience of greater bandwidth and the wider range of services (video on demand, videoconferencing, and so on) available with that greater bandwidth even if wireless QoS on the subscriber's telephone service (the primary component of the PSTN) was not as good as that of the PSTN.

Voice over
802.11

The greatest test 802.11 will have to face when being compared to the *Public Switched Telephone Network* (PSTN) is its capability to transmit voice with the clarity of the PSTN. The immediate assumption is that the *quality of service* (QoS) of an 802.11 network is inadequate for voice transmission. As demonstrated in the previous chapter, QoS on 802.11 networks can be engineered to equal that of wired networks. Its wide acceptance in enterprise networks is testimony to this. This chapter explores how voice becomes data in *voice over Internet Protocol* (VoIP) and that 802.11 is wireless IP; therefore, VoIP can travel over 802.11, delivering voice with a quality equal to that of the PSTN.

The threat to telephone companies is that it is infinitely cheaper to beam data (and voice) to customers than it is to run a copper wire or coaxial cable to them. In addition, the potential data flow to a subscriber over an 802.11 network is exponentially greater than that of the 56 Kbps delivered via a telco's copper-wire dial-up connection. The emergence of softswitch as a switching alternative to Class 4 and Class 5 switches makes it all the more feasible for 802.11 service providers to offer voice services independent of the telephone company or for subscribers (especially enterprises) to be their own telephone company, effectively bypassing the PSTN entirely.

PSTN Architecture

The PSTN, over which the vast majority of the voice traffic in North America travels, is comprised of three elements. The first is transport, or the transportation of conversation from one central office to another. The second is a telephone switch contained in the central office. A switch provides switching, or the routing of calls, in the PSTN. Finally, access denotes the connection between the switch in the central office and the subscriber's telephone or other telecommunications device. Figure 6-1 provides an overview of this architecture.

As illustrated in Figure 6-2, 802.11 is a form of access to a wider network (the PSTN, a corporate *local area network* [LAN] or the *wide area network* [WAN], or the Internet). The *Memorandum of Final Judgment* (MFJ) of 1984 opened transport to competition. The "bandwidth

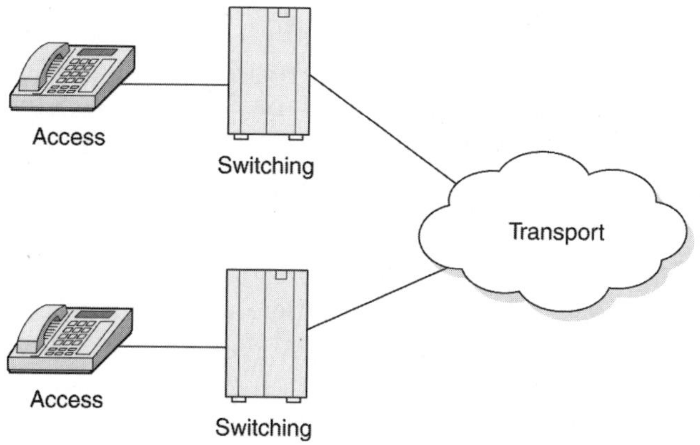

Figure 6-1
The three components of the PSTN: access, switching, and transport

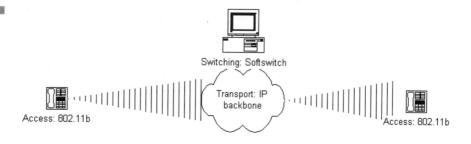

Figure 6-2
An alternative to the PSTN: 802.11 for access, softswitch for switching, and IP backbone for transport

glut" at the time of this writing has made transport relatively inexpensive. Softswitch technologies (such as IP *private branch exchange* [PBX] or Class 4 and 5 replacements) offer a viable alternative to the switching facilities of the PSTN. The Telecommunications Act of 1996 was to have opened the switching and access facilities of the PSTN to competition. For a number of reasons, this has not happened. 802.11 presents a bypass technology of the telco's copper-wire access.

Voice over Wireless: The Challenge

Wireless communications systems that are "fixed" in place have historically been divided into two different categories based on the application. The first application, delivering *Plain Old Telephone Service* (POTS) to underserved or infrastructure-challenged customers primarily in developing nations, was served by products classified as *wireless local loop* (WLL). The second application, Internet access, has to date been delivered by broadband wireless systems primarily in the enterprise market.

Because they were designed for voice with a capacity of 64 Kbps at most, the WLLs were inefficient in delivering high-speed data. *Broadband wireless access* (BWA) systems have been designed to deliver high-speed IP access first to the enterprise and, ultimately, to the residential customer.[1] They, in turn, were not designed to deliver voice. Neither of these systems is widely deployed. A wireless broadband Internet technology that also delivers good-quality voice has the potential to bypass and potentially replace the PSTN as we know it.

The emerging popularity of VoIP in the enterprise market coupled with 802.11 LANs raises the following questions: Can voice be transported over an 802.11 network? Is 802.11 not simply a wireless IP technology? If voice can travel over a wired IP network, why could it not travel over a wireless network? If transporting voice over an 802.11 network has limitations, what are they? How can they be overcome? This chapter will cover the objections to transmitting *voice over 802.11* (Vo802.11) networks and will offer solutions to those objections.

VoIP

The emergence of VoIP brings with it a wide range of possibilities. By virtue of transporting voice over a data stream, VoIP frees the voice stream from the confines of a voice-specific network and its associated platforms. VoIP can be received and transmitted via PCs, laptops, *personal digital assistants* (PDAs), and IP handsets. Where IP exists, there can be VoIP.

---

[1]"Fixed Wireless Networks and Voice," a 2002 white paper from Malibu Networks, available at www.malibu.com.

Origins of VoIP

In November 1988, Republic Telcom (yes, one *e*) of Boulder, Colorado, received patent number 4,782,485 for a "Multiplexed Digital Packet Telephone System." The plaque from the Patent and Trademark Office describes it as follows: "A method for communicating speech signals from a first location to a second location over a digital communication medium comprising the steps of: providing a speech signal of predetermined bandwidth in analog signal format at said first location; periodically sampling said speech signal at a predetermined sampling rate to provide a succession of analog signal samples; representing said analog signal samples in a digital format thereby providing a succession of binary digital samples; dividing said succession of binary digital samples into groups of binary digital samples arranged in a temporal sequence; transforming at least two of said groups of binary digital samples into corresponding frames of digital compression."

Republic and its acquiring company, Netrix Corporation, applied this voice over data technology to the data technologies of the time (X.25 and frame relay) until 1998 when Netrix and other competitors introduced VoIP onto their existing voice over data gateways. Although attempts had been made at Internet telephony from a software-only perspective, commercial applications were limited to using voice over data gateways that could interface the PSTN with data networks. Voice over data applications were popular in enterprise networks with offices spread across the globe (eliminating international interoffice long-distance bills), offices where no PSTN existed (installations for mining and oil companies), and for long-distance bypass (legitimate and illegitimate).

The popularity and applications of VoIP continued to grow. VoIP accounted for 6 percent of all international long-distance traffic in 2001.[2] Six percent may not seem like an exciting sum, but given a mere three years from the introduction of a technology to capturing 6 percent of a trillion-dollar, 100-year-old industry, it is clear that VoIP will continue to capture more market share.

---

[2]"TeleGeography 2002—Global Traffic Statistics and Commentary," TeleGeography, www.TeleGeography.com 2001.

How Does VoIP Work?

The first process in an IP voice system is the digitization of the speaker's voice. The next step (and the first step when the user is on a handset connected to a gateway using a digital PSTN connection) is typically the suppression of unwanted signals and compression of the voice signal. This has two stages. First, the system examines the recently digitized information to determine if it contains a voice signal or only ambient noise, and it discards any packets that do not contain speech. Secondly, complex algorithms are employed to reduce the amount of information that must be sent to the other party. Sophisticated *coders / decoders* (codecs) enable noise suppression and the compression of voice streams. Compression algorithms (codecs) include G.723, G.728, and G.729. G.711 is the codec for uncompressed voice at 64 Kbps.

Following compression, voice must be packetized and VoIP signaling protocols added. Some storage of data occurs during the process of collecting voice data, since the transmitter must wait for a certain amount of voice data to be collected before it is combined to form a packet and be transmitted via the network. Protocols are added to the packet to facilitate its transmission across the network. For example, each packet will need to contain the address of its destination, a sequencing number in case the packets do not arrive in the proper order, and additional data for error checking. Because IP is a protocol designed to interconnect networks of varying kinds, substantially more processing is required than in smaller networks. The network addressing system can often be very complex, requiring a process of encapsulating one packet inside another and, as data moves along, repackaging, readdressing, and reassembling the data.

When each packet arrives at the destination computer, its sequencing is checked to place the packets in the proper order. A decompression algorithm is used to restore the data to its original form, and clock-synchronization and delay-handling techniques are used to ensure proper spacing. Because data packets are transported via the network by a variety of routes, they do not arrive at their destination in order. To correct this, incoming packets are stored for a time in a jitter buffer to wait for late-arriving packets. The length of time in which data are held in the jitter buffer varies depending on the characteristics of the network.

VoIP Signaling Protocols

VoIP signaling protocols, such as H.323 and the *Session Initiation Protocol* (SIP), set up the route for the media stream or conversation over an IP network. Gateway control protocols such as the *Media Gateway Control Protocol* (MGCP) and MEGACO (also signaling protocols) establish control and status in media and signaling gateways.

Routing (*User Datagram Protocol* [UDP] and *Transmission Control Protocol* [TCP]) and transporting (*Real-Time Transport Protocol* [RTP]) the media stream (conversation) once the route of the media stream has been established are the function of routing and transport protocols. Routing protocols such as UDP and TCP could be compared to the switching function described in Chapters 2 and 3.

RTP would be analogous to the transport function in the PSTN. The signaling functions establish which route the media stream will take over the network delivering the bits, that is, the conversation. This is illustrated in Figure 6-3.

The process of setting up a VoIP call is roughly similar to that of a circuit-switched call made on the PSTN. A media gateway or IP phone must be loaded with the parameters to allow proper media encoding and the use of telephony features. Inside the media gateway is an intelligent entity known as an *endpoint*. When the calling and called parties agree on how to communicate and the signaling criteria is established, the media stream over which the packetized voice conversation will flow is established. Signaling establishes the virtual circuit over the network for that media stream. Signaling is independent of the media flow. It determines the type of media to be used in a call. Signaling is concurrent throughout the call. Two types of signaling are currently popular in VoIP: H.323 and SIP.[3]

---

[3]Bill Douskalis, *IP Telephony—The Integration of Robust VoIP Services* (New Jersey: Prentice Hall, 2000).

Figure 6-3 details the relationship between signaling and media flow. This relationship between transport and signaling is very similar to the PSTN in that *Signaling System 7* (SS7) is out-of-channel signaling, such as that used in VoIP.

H.323 H.323 is the *International Telecommunications Union—Telecommunications Standardization Sector* (ITU-T) recommendation for packet-based multimedia communication. H.323 was developed before the emergence of VoIP for video over a LAN. As it was not specifically designed for VoIP, it has faced a good deal of competition from a another protocol, SIP, which was designed specifically for VoIP over any size network. H.323 has enjoyed a first-mover advantage and a considerable installed base of H.323 VoIP networks now exists.

H.323 is comprised of a number of subprotocols. It uses protocol H.225.0 for registration, admission, status, call signaling, and control. It also uses protocol H.245 for media description and control, terminal capability exchange, and general control of the logical channel carrying the media stream(s). Other protocols make up the complete H.323 specification, which presents a protocol stack for H.323 signaling and media transport. H.323 also defines a set of call control, channel setup, and codec specifications for transmitting real-time video and voice over networks that don't offer guaranteed service or QoS. As a transport, H.323 uses RTP, an *Internet Engineering Task Force* (IETF) standard designed to handle the requirements of streaming real-time audio and video via the Internet.[4]

Figure 6-3
Signaling and transport protocols used in VoIP

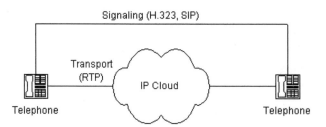

---

[4]Ibid., 9.

H.323 was the first VoIP protocol for interoperability among the early VoIP gateway/gatekeeper vendors. Unfortunately, the promise of interoperability between diverse vendors platforms did not materialize with the adoption of H.323. Given the gravity of this protocol, it will be covered in a separate chapter.

The H.323 standard is a cornerstone technology for the transmission of real-time audio, video, and data communications over packet-based networks. It specifies the components, protocols, and procedures providing multimedia communication over packet-based networks. Packet-based networks include IP-based (including the Internet) or *Internet packet exchange* (IPX)-based LANs, *enterprise networks* (ENs), *metropolitan area networks* (MANs), and WANs. H.323 can be applied in a variety of mechanisms: audio only (IP telephony); audio and video (videotelephony); audio and data; and audio, video, and data. H.323 can also be applied to multipoint-multimedia communications. H.323 provides myriad services and therefore can be applied in a wide variety of areas: consumer, business, and entertainment applications.

SIP: Alternative Softswitch Architecture? If the worldwide PSTN could be replaced overnight, the best candidate architecture, at the time of this writing, would be based on VoIP and SIP. Much of the VoIP industry has been based on offering solutions that leverage existing circuit-switched infrastructure (such as VoIP gateways that interface a PBX and an IP network). At best, these solutions offer a compromise between circuit- and packet-switching architectures with the resulting liabilities of limited features, expensive-to-maintain circuit-switched gear, and questionable QoS and reliability as a call is routed between networks based on those technologies. SIP is an architecture that potentially offers more features than a circuit-switched network.

SIP is a signaling protocol. It uses a text-based syntax similar to the *Hypertext Transfer Protocol* (HTTP) used in web addresses. Programs that are designed for parsing HTTP can be adapted easily for use with SIP. SIP addresses, known as SIP *uniform resource locators* (URLs), take the form of web addresses. A web address can be the equivalent of a telephone number in an SIP network. In addition, PSTN phone numbers can be incorporated into an SIP address for interfacing with the PSTN. An e-mail address is portable. Using the proxy concept, one can check his or her e-mail from any Internet-connected terminal in the world. Telephone numbers, simply put, are not portable. They only ring at one physical location. SIP offers a mobility function that can follow a subscriber to whatever phone he or she is nearest to at a given time.

Like H.323, SIP handles the setup, modification, and teardown of multimedia sessions, including voice. Although it works with most transport protocols, its optimal transport protocol is RTP. Figure 6-4 shows how SIP functions as a signaling protocol while RTP acts as the transport protocol for a voice conversation. SIP was designed as a part of the IETF multimedia data and control architecture. It is designed to interwork with other IETF protocols such as the *Session Description Protocol* (SDP), RTP, and the *Session Announcement Protocol* (SAP). SIP is described in the IETF's *Request for Comments* (RFC) 2543. Many in the VoIP and softswitch industry believe that SIP will replace H.323 as the standard signaling protocol for VoIP.

SIP is part of the IETF standards process and is modeled upon other Internet protocols such as the *Simple Mail Transfer Protocol* (SMTP) and HTTP. It is used to establish, change, and tear down (end) calls between one or more users in an IP-based network. In

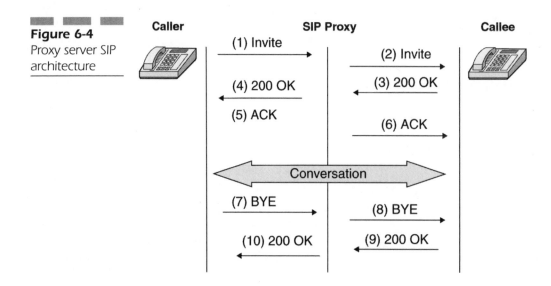

Figure 6-4
Proxy server SIP architecture

order to provide telephony services, a number of different standards and protocols must come together, specifically to ensure transport (RTP), enable signaling with the PSTN, guarantee voice quality (*Resource Reservation Setup Protocol* [RSVP]), provide directories (*Lightweight Directory Access Protocol* [LDAP]), authenticate users (*Remote Access Dial-In User Service* [RADIUS]), and scale to meet anticipated growth curves.

How Does SIP Work? SIP is focused on two classes of network entities: clients, also called *user agents* (UAs), and servers. VoIP calls on SIP to originate at a client and terminate at a server. Types of clients in the technology currently available for SIP telephony would include a personal computer loaded with a telephony agent or an SIP telephone. Clients can also reside on the same platform as a server. For example, a PC on a corporate WAN might be the server for the SIP telephony application, but it may also be used as a user's telephone (client).

SIP Architecture SIP is a client-server architecture. The client in this architecture is the UA), which interacts with the user. It usually has an interface towards the user in the form of a PC or an IP phone (an SIP phone in this case). Four types of SIP servers exist. The type of SIP server used determines the architecture of the network. The servers are UA servers, redirect servers, proxy servers, and a registrar.

Switching

In the PSTN, the switching function is performed in the central office, which contains a Class 5 switch for local calls and a Class 4 switch for long-distance calls. A Class 5 switch can cost upwards of tens of millions of dollars and is very expensive to maintain. This economy of scale has kept competitors out of the local calling market. A new technology known as softswitch is far cheaper in terms of purchase and maintenance. Potentially, softswitch enables a competitive service provider to offer its own service without having to route calls through the incumbent service provider's central office. The following pages describe softswitch.

Softswitch (a.k.a. Gatekeeper or Media Gateway Controller [MGC])

A softswitch is the intelligence in a VoIP network that coordinates the call control, signaling, and features that make a call across a network or multiple networks possible. Primarily, a softswitch performs call control. Call control performs call setups and teardowns. Once a call is set up, connection control ensures that the call stays up until the originating or terminating user releases it. Call control and service logic refer to the functions that process a call and offer telephone features. Examples of call control and service logic functions include recognizing that a party has gone off hook and that a dial tone should be provided, interpreting the dialed digits to determine where the call is to be terminated, and determining if the called party is available or busy. Such functions also include recognizing when the called party answers the phone and when either party subsequently hangs up, and recording these actions for billing.

A softswitch coordinates the routing of signaling messages between networks. Signaling coordinates actions associated with a connection to the entity at the other end of the connection. To set up a call, a common protocol must be used that defines the information in the messages and which is intelligible at each end of the network and across dissimilar networks. The main types of signaling a softswitch performs are peer to peer for call control and softswitch to gateway for media control. For signaling, the predominant protocols are SIP, SS7, and H.323. For media control, the predominant signaling protocol is MGCP.

As a point of introduction to softswitch, it is necessary to clarify the evolution to softswitch and define *Media Gateway Controller* (MGC) and gatekeeper, which were the precursors to softswitch. MGCs and gatekeepers (essentially synonymous terms for the earliest forms of softswitch) were designed to manage low-density (relative to a carrier-grade solution) voice networks. MGC communicates with both the signaling gateway and the media gateway to provide the necessary call-processing functions. The MGC uses either MGCP or MEGACO/H.248 (described in a later chapter) for intergateway communications.

Gatekeeper technology evolved out of H.323 technology (a VoIP signaling protocol described in the next chapter). As H.323 was designed for LANs, an H.323 gatekeeper can only manage activities in a zone (read LAN but not specifically a LAN). A zone is a collection of one or more gateways managed by a single gatekeeper. A gatekeeper should be thought of as a logical function, not a physical entity. The functions of a gatekeeper are address translation (that is, a name or e-mail address for a terminal or gateway and a transport address) and admissions control (it authorizes access to the network).

As VoIP networks got larger and more complex, management solutions with far greater intelligence became necessary. Greater call-processing power became necessary, as did the ability to interface signaling between IP networks with the PSTN (VoIP signaling protocols to SS7). Other drivers included the need to integrate features on the network and interface disparate VoIP protocols. Thus, softswitch was born.

A significant market driver for softswitch is the protocol intermediation necessary to interface H.323 and SIP networks, for example. Another market driver for softswitch is to interface between the PSTN (SS7) and IP networks (SIP and H.323). Another function for softswitch is the intermediation between media gateways of dissimilar vendors. Despite an emphasis on standards such as H.323, interoperability remains elusive. A softswitch application can overcome intermediation issues between media gateways. More information on VoIP protocols and signaling IP to PSTN is provided in following chapters.

The softswitch provides usage statistics to coordinate billing, track operations, and administrative functions of the platform while interfacing with an application server to deliver value-added subscriber services. The softswitch controls the number and type of features provided. It interfaces with the application server to coordinate features (conferencing, call forwarding, and so on) for a call.

Physically, a softswitch is software hosted on a server chassis filled with IP boards and it includes the call control applications and drivers.[5] Very simply, the more powerful the server, the more capable the softswitch. That server need not be co-located with other components of the softswitch architecture.

Other Softswitch Components

The key advantage of softswitch over its circuit-switched predecessor is that it utilizes distributed architecture. That is, its components do need to be co-located. Those components include the signaling gateway, media gateway, and application server (see Figure 6-5).

Signaling Gateway Signaling gateways are used to terminate signaling links from PSTN networks or other signaling points. The SS7 signaling gateway serves as a protocol mediator (translator) between the PSTN and IP networks. That is, when a call originates in an IP network using H.323 as a VoIP protocol and must terminate in the PSTN, a translation from the H.323 signaling protocol to SS7 is necessary in order to complete the call. Physically, signaling function can be embedded directly into the MGC or housed within a standalone gateway.

Figure 6-5
Relationship of
softswitch
components

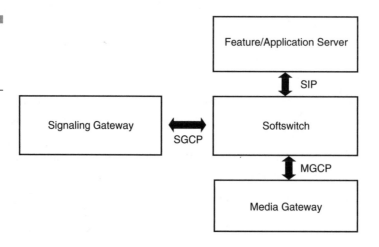

Media Gateway The media gateway converts an analog or circuit-switched voice stream to a packetized voice stream. Media gateways rank from one- or two-port residential gateways to carrier-grade platforms with 100,000 ports. The media gateway can be located at the customer's premises or co-located at the central office.

Application Server The application server accommodates the service and feature applications made available to a service provider's customers. Examples include call forwarding, conferencing, voice mail, forward on busy, and so on. Physically, an application server is a server loaded with a software suite that offers the application programs. The softswitch accesses these and then enables and applies them to the appropriate subscribers as needed.

A softswitch solution emphasizes open standards as opposed to the Class 4 or 5 switch that historically offered a proprietary and closed environment. A carrier was a "Nortel shop" or a "Lucent shop." No components (hardware or software) from one vendor would be compatible with products from another vendor. Any application or feature on a DMS-250, for example, had to be a Nortel product or specifically approved by Nortel. This usually translates into less than competitive pricing for those components. Softswitch open standards are aimed at freeing service providers from vendor dependence and the long and expensive service development cycles of legacy switch manufacturers.

Concern exists among service providers as to whether a softswitch solution can transmit a robust feature list identical to those found on a 5ESS Class 5 switch, for example. Softswitch offers the advantage of allowing a service provider to integrate third-party applications or even write their own while interoperating with the features of the PSTN via SS7. This is potentially the greatest advantage to a service provider presented by softswitch technology.

Features reside at the application layer in softswitch architecture. The interface between the Call Control Layer and specific applications is the *Application Program Interface* (API). Writing and interfacing an application with the rest of the softswitch architecture occurs in the Service Creation Environment.

VoIP and Softswitch Pave the Way for Vo802.11

A number of attempts have been made to deploy voice services via WLL, that is, using wireless technologies (not 802.11) to offer telephone service in underserved or third-world markets. As these services have been limited to voice services, they have not caught on with a mass market. 802.11 is different in that it is a protocol for Ethernet over a wireless medium. The building blocks for a potential alternative to the PSTN now fall into place. Not only does an 802.11 network offer a potential bypass of the PSTN for voice services, it also offers broadband Internet and its incumbent suite of services.

Objections to VoIP over 802.11 (Vo802.11)

Just like concerns over 802.11 as a whole, five major objections arise in adopting VoIP over 802.11 (Vo802.11): voice quality as it relates to QoS, security, E911 (Emergency 911), *Communications Assistance for Law Enforcement Act* (CALEA), and range. Many suspect that an adverse tradeoff takes place between the predictability (QoS) of copper wires that deliver voice to residences and the unpredictability of the airwaves as utilized by 802.11. As regards security, some fear their conversations could be susceptible to eavesdropping if their conversation is carried in the open air. Fortunately, the industry now offers solutions to make Vo802.11 as good quality-wise and as secure as the PSTN. Many will raise objections regarding requirements for E911 and CALEA for carriers. Solutions for range issues related to 802.11 are contained in Chapter 3, "Range Is Not an Issue," and will not be covered here as no voice-specific solutions have been made regarding the range of 802.11. The following pages explain solutions for these objections to Vo802.11. No problems exist regarding Vo802.11, only solutions.

Objection 1: Voice Quality of Vo802.11

Despite the fact that telephone companies are losing thousands of lines per month in the United States to cell phone service providers, many perceive that voice over a cell phone connection would deliver inferior voice quality and, as a result, is not a viable alternative to the copper wires of the PSTN. As explored in the previous chapter, a number of new measures (primarily 802.11e) improve QoS on 802.11.

But what about voice? As wired service providers and network administrators have found, voice is the hardest service to provision on an IP network. New developments in the Vo802.11 industry point to some exciting developments that overcome the chief objection to Vo802.11. First, it is necessary to determine which metrics to use in comparing Vo802.11 to the voice quality of the PSTN.

Measuring Voice Quality in Vo802.11   How does one measure the difference in voice quality between a Vo802.11 and the PSTN? As the VoIP industry matures, new means of measuring voice quality are arriving on the market. Currently, two tests award some semblance of a score for voice quality. The first is a holdover from the circuit-switched voice industry known as *Mean Opinion Score* (MOS). The other has emerged with the rise in popularity of VoIP and is known as *Perceptual Speech Quality Measurement* (PSQM).

Mean Opinion Score (MOS)   Can voice quality as a function of QoS be measured scientifically? The telephone industry employs a subjective rating system known as the MOS to measure the quality of the telephone connections. The measurement techniques are defined in ITU-T P.800 and are based on the opinions of many testing volunteers who listen to a sample of voice traffic and rate the quality of that transmission. The volunteers listen to a variety of voice samples and are asked to consider factors such as loss, circuit noise, side tone, talker echo, distortion, delay, and other transmission problems. The volunteers then rate the voice samples from 1 to 5, with 5 being excellent and 1 being bad. The voice samples are then awarded an MOS. An MOS of 4 is considered toll quality.

It should be stated here that the voice quality of VoIP applications can be engineered to be as good or better than the PSTN. Recent research performed by the Institute for Telecommunications Sciences in Boulder, Colorado, compared the voice quality of traffic routed through VoIP gateways with the PSTN. Researchers were fed a variety of voice samples and were asked to determine if the sample originated with the PSTN or from the VoIP gateway traffic. The result of the test was that the voice quality of the VoIP-gateway-routed traffic was "indistinguishable from the PSTN."[6] It should be noted that the IP network used in this test was a closed network and not the public Internet or another long-distance IP network. This report indicates that quality media gateways can deliver voice quality on the same level as the PSTN. The challenge then shifts to ensuring the IP network can deliver similar QoS to ensure good voice quality. This chapter explains how measures can be taken to engineer voice-specific solutions into a wireless network to ensure voice quality equal to the PSTN.

Perceptual Speech Quality Measurement (PSQM) Another means of testing voice quality is known as PSQM. It is based on ITU-T Recommendation P.861, which specifies a model for mapping actual audio signals to their representations inside the head of a human. Voice quality consists of a mix of objective and subjective parts, and it varies widely among the different coding schemes and the types of network topologies used for transport. In PSQM, a measurement of processed signals (compressed, encoded, and so on) derived from a speech sample are collected, and an objective analysis is performed, comparing the original and the processed version of the speech sample. From that, an opinion is rendered as to the quality of the signal-processing functions that processed the original signal. Unlike MOS scores, PSQM scores result in an absolute number, not a relative comparison between the two signals.[7] The value in this is that vendors can state the PSQM score for a given platform (as

---

[6]Andrew Craig, "Qualms of Quality Dog Growth of IP Telephony," *Network News* (November 11, 1999): 3.

[7]Douskalis, 242–243.

assigned by an impartial testing agency). Service providers can then make at least part of their buying decision based on the PSQM score of the platform (see Figure 6-6).

Detractors to Voice Quality in 802.11 What specifically detracts from good voice quality in an 802.11 environment? Latency, jitter, packet loss, and echo all affect voice quality on an 802.11 network. With proper engineering, the impact of these factors on voice quality can be minimized, and voice quality equal to or better than that of the PSTN can be achieved on 802.11 networks.

Latency (Also Known as Delay) Voice as a wireless IP application presents unique challenges for 802.11 networks. Primary among these is acceptable audio quality resulting from minimized network delay in a mixed voice and data environment. Ethernet, wired or wireless, is not designed for real-time streaming media or guaranteed packet delivery. Congestion on the wireless network, without traffic differentiation, can quickly render voice unusable. QoS measures must be taken to ensure voice packet delays stay under 100 milliseconds.

Voice signal processing at the sending and receiving ends that includes the time required to encode or decode the voice signal from analog or digital form into the voice-coding scheme selected for the call and vice versa adds to the delay. Compressing the voice signal also increases the delay. The greater the compression, the greater the delay. Where bandwidth costs are not a concern, a service provider can utilize G.711, which is uncompressed voice (64 Kbps) and imposes a minimum of delay due to the lack of compression.

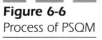

Figure 6-6
Process of PSQM

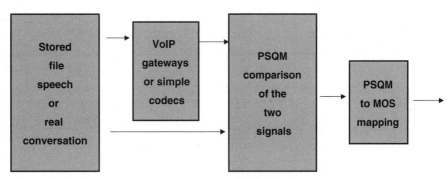

On the transmit side, packetization delay is another factor that must be entered into the calculations. The packetization delay is the time it takes to fill a packet with data. The larger the packet size, the more time is required. Using shorter packet sizes can shorten this delay but will increase the overhead because more packets have to be sent, all containing similar information in the header. Balancing voice quality, packetization delay, and bandwidth utilization efficiency is very important to the service provider.[8]

How much delay is too much? Of all the factors discussed in Chapter 4, "Security and 802.11," that degrade VoIP, latency (delay) is the greatest. Recent testing by Mier Labs offers a metric for VoIP voice quality. That measure is to determine how much latency is acceptable comparable to the voice quality offered by the PSTN. Latency less than 100 *milliseconds* (ms) does not affect toll-quality voice. However, latency over 120 ms is discernable to most callers, and at 150 ms the voice quality is noticeably impaired, resulting in less than toll-quality communication. The challenge for VoIP service providers and their vendors is to get the latency of any conversation on their network to not exceed 100 ms.[9] Humans are intolerant of speech delays of more than about 200 ms. As mentioned earlier, ITU-T G.114 specifies that delay is not to exceed 150 ms one-way or 300 ms round trip. The dilemma is that although elastic applications (e-mail, for example) can tolerate a fair amount of delay, they usually try to consume every bit of network capacity they can. In contrast, voice applications need only small amounts of the network, but that amount has to be available immediately.[10]

The delay experienced in a call occurs on the transmitting side, on the network, and on the receiving side. Most of the delay on the transmitting side is due to codec delay (packetization and lookahead) and processing delay. On the network, most of the delay

---

[8]Douskalis, 230–231.

[9]Mier Communications, "Lab Report-QoS Solutions," www.sitaranetworks.com/solutions/pdfs/mier_report.pdf February 2001.

[10]John McCullough and Daniel Walker, "Interested in VoIP? How to Proceed," *Business Communications Review* (April 1999): 16–22.

stems from transmission time (serialization and propagation) and router queuing time. Finally, the jitter buffer depth, processing, and, in some implementations, the polling intervals add to the delay on the receiving side.

The delay introduced by the speech coder can be divided into algorithmic and processing delays. The algorithmic delay occurs due to framing for block processing, since the encoder produces a set of bits representing a block of speech samples. Furthermore, many coders using block processing also have a look-ahead function that requires a buffering of future speech samples before a block is encoded. This adds to the algorithmic delay. Processing delay is the amount of time it takes to encode and decode a block of speech samples.

Dropped Packets   In IP networks, a percentage of the packets can be lost or delayed, especially in periods of congestion. Also, some packets are discarded due to errors that occurred during transmission. Lost, delayed, and damaged packets result in a substantial deterioration of voice quality. In conventional error-correction techniques used in other protocols, incoming blocks of data containing errors are discarded, and the receiving computer requests the retransmission of the packet. Thus, the message that is finally delivered to the user is exactly the same as the message that originated. As VoIP (and tangentially Vo802.11) systems are time sensitive and cannot wait for retransmission, more sophisticated error-detection and correction systems are used to create sound to fill in the gaps. This process stores a portion of the incoming speaker's voice and then, using a complex algorithm to approximate the contents of the missing packets, new sound information is created to enhance the communication. Thus, the sound heard by the receiver is not exactly the sound transmitted, but rather portions of it have been created by the system to enhance the delivered sound.[11]

Most of the packet losses occur in the routers, either due to a high router load or a high link load. In both situations, packets in the

---

[11]Report to Congress on Universal Service, CC Docket No. 96-45, a white paper on IP voice services, March 18, 1998.

queues might be dropped. Another source of packet loss is errors in the transmission links, resulting in *cyclic redundancy check* (CRC) errors for the packet. Configuration errors and collisions might also result in packet losses. In non-real-time applications, packet losses are solved at the protocol layer by retransmission (TCP). For telephony, this is not a viable solution since retransmitted packets would arrive too late and be of no use.

Perhaps the chief challenge to Vo802.11 is that, relative to wired networks, packets are dropped at an excessive rate (upwards of 30 percent). This can lead to distortion of the voice to the extent that the conversation is unintelligible. In VoIP gateways designed for wired networks, one solution is to use a jitter buffer with a "bit bucket." The solution in the wired VoIP industry had been to simply eliminate (drop) voice packets that arrive late and out of order. This is acceptable if the percentage of late and out-of-order packets is fairly small (say, less than 10 percent). When the packet loss grows due to the many vagaries of wireless transmissions, the voice quality falls off precipitously.

Jitter   Jitter occurs because packets have varying transmission times. It is caused by different queuing times in the routers and possible different routing paths. The jitter results in unequal time spacing between the arriving packets, and it requires a jitter buffer to ensure a smooth, continuous playback of the voice stream.

The chief correction for jitter is to include an adaptive jitter buffer. The jitter buffer described in the solution earlier is a fixed jitter buffer. The improvement is an adaptive jitter buffer that can dynamically adjust to accommodate for high levels of delay encountered in wireless networks.

A Word About Bit Rate (or Compression Rate)   The bit rate is the number of bits per second delivered by the speech encoder and therefore determines the bandwidth load on the network. It is important to note that the packet headers (IP, UDP, and RTP) also add to the bandwidth. Speech quality generally increases with the bit rate. Very simply put, the greater the bandwidth, the greater the speech quality.

Solution: Voice Codecs Designed for VoIP, Especially VoIP over 802.11

Many of the detractors to good speech quality in VoIP over 802.11 can be overcome by engineering a variety of fixes into the speech codecs used in both circuit- and packet-switched telephony. The following pages will describe speech coding and how it applies to speech quality.

Speech Coding

Voice or speech compression, as performed by speech codecs, is a crucial factor affecting the speech quality, and thereby QoS, in IP telephony. The speech codecs are deployed at the endpoints (gateways, IP phones, or PCs) and therefore determine the achievable end-to-end quality. The speech encoder converts the digitized (after analog-to-digital conversion) speech signal to a bitstream. This bitstream is packetized and transported over the IP network. The speech decoder reconstructs the speech signal from the bits in the received packets. The reconstructed speech signal is an approximation of the original signal. A speech codec has several attributes, and the most important ones are outlined in Table 6-1.

Modifying Voice Codecs to Improve Voice Quality

One of the first processes in the transmission of a telephone call is the conversion of an analog signal (the wave of the voice entering the telephone) into a digital signal. This process is called *pulse code modulation* (PCM). This is a four-step process consisting of *pulse amplitude modulation* (PAM) sampling, companding, quantization, and encoding. Encoding is a critical process in VoIP and Vo802.11. To date, voice codecs used in VoIP (packet switching) are taken directly from PSTN

Table 6-1

Popular voice codecs developed for circuit-switched telephony

ITU Standard	Description
P.800	Subjective rating system to determine the MOS or the quality of telephone connections
G.114	Maximum one-way delay end-to-end for a VoIP call (150 ms)
G.165	Echo cancellers
G.168	Digital network echo cancellers
G.711	PCM of voice frequencies
G.722	7 kHz audio coding within 64 Kgps
G.723.1	Dual-rate speech coder for multimedia communications transmitting at 5.3 and 6.3 Kgps
G.729	Coding for speech at 8 Kgps using *conjugate structure algebraic code excited linear prediction* (CS-ACELP)
G.729A	Annex A reduced complexity 8 Kgps CS-ACELP speech codec
H.323	Packet-based multimedia communications system
P.861	Specifies a model to map actual audio signals to their representations inside the human head
Q.931	*Digital Subscriber Signaling System* (DSS) No. 1 *Integrated Services Digital Network* (ISDN) User-Network Interface Layer 3 Specification for Basic Call Control

technologies (circuit switching). Cell phone technologies use PSTN voice codecs. New software in the Vo802.11 industry utilizes modified PSTN codecs to deliver voice quality comparable to the PSTN.

Encoding The final process in the PCM process used in circuit switching (as opposed to packet switching in IP networks) is encoding the voice signal. This is performed by a codec, of which three types exist: waveform codecs, source codecs (also known as vocoders), and hybrid codecs.

Waveform codecs sample and code an incoming analog signal without regard to how the signal was generated. Quantized values of the samples are then transmitted to the destination where the original signal is reconstructed at least to a certain approximation of the original. Waveform codecs are known for simplicity with high-quality output. The disadvantage of waveform codecs is that they consume considerably more bandwidth than the other codecs. When waveform codecs are used at a low bandwidth, speech quality degrades markedly.

Source codecs match an incoming signal to a mathematical model of how speech is produced. They use the linear predictive filter model of the vocal tract, with a voiced/unvoiced flag to represent the excitation that is applied to the filter. The filter represents the vocal tract and the voice/unvoiced flag represents whether a voiced or unvoiced input is received from the vocal chords. The information transmitted is a set of model parameters as opposed to the signal itself. The receiver, using the same modeling technique in reverse, reconstructs the values received into an analog signal.

Source codecs operate at low bit rates and reproduce a synthetically sounding voice. Using higher bit rates with source codecs does not result in improved voice quality. Vocoders (source codecs) are most widely used in private and military applications. Hybrid codecs are deployed in an attempt to derive the benefits from both technologies. They perform some degree of waveform matching while mimicking the architecture of human speech. Hybrid codecs provide better voice quality at a low bandwidth than waveform codecs. The following section examines the popular speech codecs.

G.711   G.711 is the best known coding technique in use today. It is the coding technique used in circuit-switched telephone networks all over the world. G.711 has a sampling rate of 8,000 Hz. If uniform quantization were to be used, the signal levels commonly found in speech would be such that at least 12 bits per sample would be needed, giving it a bit rate of 96 Kbps. Nonuniform quantization is used with eight bits to represent each sample. This quantization leads to the well-known 64 Kbps DS0 rate. G.711 is often referred to as PCM.

G.711 has two variants: A-law and mu-law. Mu-law is used in North America and Japan where T-carrier systems prevail. A-law is used everywhere else in the world. The difference between the two is the way nonuniform quantization is performed. Both are symmetrical at approximately zero. Both A-law and mu-law offer good voice quality with an MOS of 4.3. Despite being the predominant codec in the industry, G.711 suffers one significant drawback: It consumes 64 Kbps in bandwidth. Carriers seek to deliver like-voice quality using less bandwidth, thus saving on operating costs.

G.723.1 ACELP G.723.1 ACELP can operate at either 6.3 or 5.3 Kbps with the 6.3 Kbps providing higher voice quality. Bit rates are contained in the coder and decoder, and the transition between the two can be made during a conversation. The coder takes a bank-limited input speech signal that is sampled at 8,000 Hz and undergoes uniform PCM quantization, resulting in a 16-bit PCM signal. The encoder then operates on blocks or frames of 240 samples at a time. Each frame corresponds to 30 ms of speech, which means that the coder causes a delay of 30 ms. Including a look-ahead delay of 7.5 ms gives a total algorithmic delay of 37.5 ms.

G.723.1 gives an MOS of 3.8 in a circuit-switched application, which is highly advantageous in regards to the bandwidth used. The delay of 37.5 ms one way presents an impediment to good quality, but the round-trip delay over varying aspects of a network determines the final delay and not necessarily the codec used.

G.729 G.729 is a speech coder that operates at 8 Kbps. This coder uses input frames of 10 ms, corresponding to 80 samples at a sampling rate of 8,000 Hz. This coder includes a 5 ms look-ahead, resulting in an algorithmic delay of 15 ms (considerably better than G.723.1). G.729 uses an 80-bit frame. The transmitted bit rate is 8 Kbps. Given that it turns in an MOS of 4.0, G.729 is perhaps the best tradeoff in bandwidth for voice quality.

The previous paragraphs provide an overview of the multiple means of maximizing the efficiency of transport via the PSTN. What we find today is that *time-division multiplexing* (TDM) is synonymous with circuit switching. Telecommunications engineers use the term TDM to describe a circuit-switched solution. The standard in use on the PSTN is 64 Kbps.

The codecs described in the previous pages apply to VoIP as well. VoIP engineers seeking to squeeze more conversations over valuable bandwidth have found these codecs very valuable in compressing VoIP conversations over an IP circuit.[12]

The construction of IP packets to be transmitted determines both the bandwidth and delay. Smaller packets, in terms of the number of speech bits, are less efficient since the overhead for header information is the same as for larger packets. Furthermore, the delay increases when several speech-coded frames are sent in the same packet. Table 6-2 shows the actual bandwidth load on the network, including header information. The numbers are based on the common situation where IP, UDP, and RTP (40 bytes per packet) are used. All numbers are based on 20 ms packets, except for G.723.1 where the 30 ms frame size requires 30 ms packets. The delay figure given includes encoding delay but not processing delay and packet-assembling delay. It can be noted that with voice activity detection and silence compression, the bandwidth load is reduced by approximately 50 percent. It is evident from the table

Table 6-2

MOS scores of speech codecs

Standard	Data Rate (Kbps)	Delay (ms)	MOS
G.711 G.721 G.723	64	0.125	4.8
G.726	16, 24, 32, and 40	0.125	4.2
G.728	16	2.5	4.2
G.729	8	10	4.2
G.723.1	5.3 and 6.3	30	3.5 and 3.98

[12]Ibid.

that for the low-rate codecs, headers contribute to the major part of the bandwidth.

Circuit-Switched Speech Coding in IP Telephony   The most commonly used codecs for IP telephony today are G.711, G.729, and G.723.1 (at 6.3 Kbps). All these codecs were designed for, or based on technology designed for, circuit-switched telephony. Mobile telephony has been the major driver for the development of speech-coding technology in recent years. All the coders used in mobile telephony, as well as G.729 and G.723.1, are based on the CELP paradigm. These codecs are designed for use in circuit-switched networks and do not work well for packet-switched networks, as their design is focused on handling bit errors rather than packet losses. The important points regarding G.711 as a Vo802.11 codec are that the coder was designed for circuit-switched telephony and it does not include any means to counter packet loss. The insertion of zeros is commonly used when packet loss occurs, leading to a disrupted voice stream (coming in "broken") and the steep degradation of quality with increasing packet losses.

It is possible to introduce error concealment by extrapolating and interpolating received speech segments, which improves quality. An example is the new Annex I to G.711 called G.711 PLC, which does not always work well and does not guarantee robust operation.

G.729 and G.723.1 belong to a different class of coders compared to G.711. Many important points must be made regarding G.729 and G.723.1 (as well as other CELP coders). The coding paradigm used in these coders has been developed for circuit-switched and mobile telephony. The basic speech quality is worse than PSTN quality (that is, they have mobile telephony quality) and the coding process is based on interframe dependencies leading to interpacket dependencies. Packet loss performance is also very poor because of error propagation resulting from interpacket dependencies and speech quality degrades rapidly with increasing packet losses. The coders have built-in heuristic error concealment methods and they also suffer from interframe dependencies (for some coders, more frames than the lost one need error concealment). The coders also produce an inflexible bitstream and the packet size is restricted to an integer number of frames, which reduces flexibility.

The QoS Solution: Fix Circuit-Switched Voice Codecs in a Packet-Switched, Wireless World with Enhanced Speech-Processing Software

If circuit-switching voice codecs are the challenge to good QoS in wireless, packet-switched networks, what then is the fix for outdated voice codecs? A market of enhanced speech-processing software is emerging that corrects for the shortcomings of traditional voice codecs that were designed decades ago for a circuit-switched PSTN. These recent developments in Vo802.11b software provide QoS enhancement solutions for IP telephony in the terminal with very high voice quality even with severe network degradations caused by jitter and packet loss. These Vo802.11b QoS enhancements should provide Vo802.11b speech quality comparable to that of the PSTN. Also, speech quality should degrade gradually as packet loss increases. Moderate packet loss percentages should be inaudible.

Enhanced Speech-Processing Software

New speech-processing algorithms provide for diversity, which means that an entire speech segment is not lost when a single packet is lost. Diversity is achieved by reorganizing the representation of the speech signal. Diversity does not add redundancy or send the same information twice. Ergo, it is bandwidth efficient and ensures that packet losses lead to a gradual and imperceptible degradation of voice quality. The tradeoff is that diversity leads to increased delays. Enhanced speech-processing software includes advanced signal processing to dynamically minimize delay. Therefore, the overall delay is maintained at approximately the same level as it would be without diversity. Furthermore, the basic quality (no packet loss) is equivalent to or better than PSTN (using G.711).

Enhanced speech-processing software is built to enhance existing standards used in IP telephony. This software enables high speech quality on a loaded network with jitter, high packet losses, and delays. Cost savings are realized using enhanced speech-processing software, as no need exists to overprovision network infrastructure. The high packet loss tolerance also reduces the need for and subsequent cost of network supervision, resulting in further cost savings.

Examples of Enhanced Speech-Processing Products

Voice Quality in Vo802.11 can be made to equate that of the PSTN by implementing a number of measures. The following pages will explain those measures and how they improve voice quality.

Adaptive Jitter Buffer   The use of an adaptive jitter buffer optimizes sound quality by using an advanced adaptive jitter buffer control combined with an error concealment algorithm. One such product is GIPS NetEQ™ from San Francisco-based Global IP Sound. It works with any codec such as iLBC (GIPS low bit rate codec), G.711 (including GIPS Enhanced G.711), G.729, and G.723.1. NetEQ is only required on the receiving end of any conversation. NetEQ improves the sound quality significantly without any interoperability problems. This solution quickly adapts to the dynamic network conditions of packet-switched networks. This ensures high speech quality with significant latency savings compared to conventional jitter-buffering technology.

Enhanced G.711   G.711 with enhancement provides superior packet loss robustness. Enhanced G.711 consists of the G.711 codec combined with an enhancement to provide packet loss robustness. During the call setup, it is determined if the recipient also has Enhanced G.711. If so, the call will continue using GIPS Enhanced G.711; if there is not a match, the call will proceed using G.711 on both ends. The enhancement unit is similar in function to encryption methods. The packets are transcoded to prevent packet loss as opposed to privacy. Enhanced G.711 in combination with an adaptive jitter buffer provides a PSTN speech-quality level at packet loss/delay rates up to 30 percent. This is achieved without increasing the bit rate and without increasing latency significantly.

Packet Loss Robustness   Using a low bit rate codec is one method for increasing packet loss robustness; low bit rate codecs use less bandwidth, providing a more efficient use of the available bandwidth. The basic speech quality of one low bit rate codec, GIPS iLBC freeware, offers better voice quality than G.729 and G.723.1, and it operates at a rate of 13.3 Kbps. Another method of increasing robustness to packet loss is to use an error concealment algorithm, such as the aforementioned GIPS NetEQ.

Acoustic Echo Cancellation   Echo is often prevalent when using a PC or IP phone. Acoustic echo cancellation is contained in enhanced speech software to counter echo in those platforms.

Results of Enhanced Speech Software: An Independent Evaluation by Lockheed Martin

Lockheed Martin Global Telecommunications (LMGT) recently performed independent subjective tests of speech enhancement software products from San Francisco-based Global IP Sound. The tests show that GIPS speech and audio-processing software NetEQ, GIPS Enhanced G.711, and iPCM-wb outperform the legacy VoIP enhancement products in their respective categories. When installed in the IP edge devices, such as media gateways and IP phones, GIPS solutions give equipment and network service providers outstanding QoS even during peak traffic hours.

The independent test laboratory within LMGT (previously a part of COMSAT Laboratories) conducted the subjective testing of the GIPS sound-processing software. The tests compared conventional speech codecs with GIPS NetEQ, adaptive jitter buffer and error concealment, GIPS Enhanced G.711, a standard-improvement solution, and iPCM-wb, a high-end (wideband) speech processing codec. The LMGT Test Report concluded, "The perceived speech quality performances of the GIPS telecommunications and wideband solutions show a dramatic improvement of that of traditional G.711 for the range of impaired-channel scenarios tested."

The test results reveal that even under situations with high packet loss and delay, Global IP Sound's enhanced speech-processing software for telephony bandwidth delivers a sound quality that matches or is very close to PSTN quality. The GIPS wideband speech solutions for end-to-end IP applications deliver higher than PSTN quality with very high robustness to network degradations.

The test results verify that NetEQ, which is codec independent, significantly improves the sound quality of the standard G.711 codec under conditions with packet loss/delay. The combination of GIPS Enhanced G.711 and NetEQ delivers sound quality matching PSTN for moderate packet loss/delay levels while being very close to PSTN for high packet loss/delay. The combination of iPCM-wb and NetEQ-wb (products aimed at the broadband market) offers high-fidelity sound quality even under adverse network conditions with up to 30 percent packet loss/delay (see Figure 6-7).

The methodologies and test listening groups were selected in accordance with acknowledged criteria set out by the ITU, ETSI, and the TIA, the relevant international standards bodies.[13] The conclusion of these tests is that, given the vagaries of both VoIP and wireless environments, enhanced speech software can correct most adverse conditions in Vo802.11 regarding latency, jitter, and packet loss and still deliver a voice quality that is comparable to that of the PSTN.

Figure 6-7
Results of Lockheed Martin testing of voice quality using Global IP Sound-enhanced speech software products. Note that a voice quality comparable to the PSTN is achieved.
Source: Global IP Sound

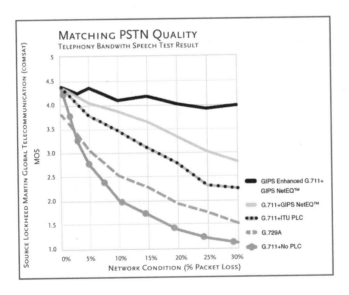

[13]Third-party Evaluation of GLOBAL IP SOUND Edge Device QoS Solutions for VoIP, a white paper from Global IP Sound, available online at www.globalipsound.com.

Objection 2: Security for Vo802.11

Although a chapter was devoted earlier in this book to security on 802.11 networks, it is important to examine how security applies specifically to Vo802.11. The misperception is that, because the conversation is transmitted over the airwaves, the voice stream is susceptible to interception, that is, eavesdropping. Although such an occurrence is not entirely impossible, it would be extremely difficult to tap into such a conversation.

Case Study for Security of Vo802.11: SpectraLink's Secure Radio Technology Vo802.11 telephone systems provide additional measures of security through sophisticated radio technology and proprietary signal encoding. Many Fortune 500 companies use Vo802.11 telephone systems in their most secure areas, such as executive offices, data centers, and network control centers. Vo802.11 telephone systems employ digital spread spectrum transmission and a pseudorandom hopping sequence against radio eavesdropping.

Digital Spread Spectrum Transmission The Vo802.11 telephone systems use a proprietary implementation of frequency hopping spread spectrum radio transmission, a radio technology originally developed by the military for secure and covert communications. Spread spectrum takes a discrete signal, such as a digitized voice conversation, and spreads it over a wide range of frequencies rather than transmitting at a single carrier frequency. Vo802.11 phone systems use frequency hopping to spread the signal by changing the carrier frequency once every 10 milliseconds (100 times every second). Because the carrier frequency changes rapidly, a radio scanner or narrowband receiver cannot be used to recover the information. These systems use 25 to 50 different frequencies in the hopping sequence, so a narrowband scanner has access to less than 4 percent of any conversation.

Proprietary Pseudorandom Hopping Sequence A proprietary pseudorandom hopping sequence provides additional security. If a potential eavesdropper went to the expense of developing a scanner that could change frequencies 100 times a second, he or she would then have to attempt to determine the pseudorandom sequence to know when to monitor which frequency. Furthermore, each base station is transmitting on a different frequency, so multiple scanners would be required to follow a conversation as it was handed off from base station to base station.

Vo802.11 systems use digital transmission, meaning that the analog voice signal is converted to a digital signal. This digital signal is scrambled to improve transmission, further complicating the ability to interpret an intercepted signal. Finally, Vo802.11 systems use *Time Division Multiple Access* (TDMA) to provide multiple speech channels from a single base station. The frame format and signaling for a Vo802.11 TDMA signal are proprietary and would have to be determined to identify discrete conversations on the radio link.[14]

Objection 3: CALEA and E911

An objection to VoIP often posed by the circuit-switched industry concerns a telecommunications carrier's compliance with CALEA and E911, which are legal requirements for primary-line telephone service providers in the United States. The laws requiring telephone companies to provide these services were made before the Internet became part of mainstream America. Although the technological means for providing these services in a number of circumstances exists, the first question that should be asked is what obligation does the service provider have in providing these services?

802.11 is, at the time of this writing, primarily used in enterprise environments. An enterprise has no obligation to provide itself with CALEA or E911 services. Where these requirements may gain greater scrutiny is in residential applications along the lines of providing "lifeline services." 802.11 Internet access is provided by *wireless Internet service providers* (WISPs). The *Federal Communications Commission* (FCC) considers ISPs (and WISPs) to be information service providers, not telecommunications service providers. The dis-

---

[14]Geri Mitchell, "Radio Security of the Link Wireless Telephone System," a white paper from SpectraLink, www.spectralink.com.

tinction between them is that information service providers are *not* required to provide the same scope of services required of telecommunication service providers. Therefore, a WISP is under no obligation to provide CALEA or E911 services.

WISPs are also considered to be second-line service providers, as opposed to primary-line service providers. Primary-line service providers are considered telecommunications service providers and are expected to comply with the same laws that incumbent telephone companies must comply with. A short list of those requirements includes CALEA, E911, paying into the Universal Service Fund, and paying access fees when originating or terminating long-distance calls.

Finally, a WISP may have little or no knowledge of its subscribers using Vo802.11. No unique equipment or recording is required of the WISP for voice calls to originate or terminate on its network. All a WISP does is provide access to an IP network (namely the Internet or a managed IP network). The subscribers then equip themselves with IP handsets, Vo802.11 phones, VoIP software on their PCs, and so on. In such a scenario, the subscriber is not required to provide themselves with CALEA or E911.

E911 A number of E911 solutions are arriving on the market at the time of this writing. First, some solutions are overflowing from the cell phone industry where E911 will soon be a requirement. One solution in the case of Vo802.11 is a mechanism to triangulate the signals emanating from the victim's 802.11 handset. Different access points can report the vector and distance to the subject Vo802.11 handset. Another solution is to include *Global Positioning Satellite* (GPS) technology in a Vo802.11 handset. That way the exact location of the handset is known at any time.

CALEA This requirement may be relaxed in a forthcoming regulatory regime outlined by FCC Chairman Michael Powell in an October 30, 2002 address at University of Colorado. In that speech, Chairman Powell conceded that the CALEA law was designed for the circuit-switched world and was, at the time of that speech, almost impossible to comply with in a Vo802.11 environment. As a result, and in the interest of promoting all that 802.11 and similar technologies have to offer, Chairman Powell hinted that such requirements would have to be relaxed.

It should also be emphasized that CALEA and E911 are requirements for the U.S. market. Vendors would do well to not focus solely on the U.S. market and let these requirements for the circuit-switched world restrict them in developing products and services that would benefit consumers around the world.

Architecture of Vo802.11: Putting It All Together

What is the architecture for an alternative to the PSTN? The PSTN is comprised of three elements: access (the wires to a residence, for example), switching (the switches in the central office), and transport (long-distance ATM networks or IP fiber-optic backbones). Figure 6-2 shows how voice services are handled via an alternative network where access is performed by 802.11 (or associated protocols), switching is done by a softswitch, and transport is handled by IP fiber-optic backbone. The common denominator in this alternative to the PSTN is VoIP. Anywhere access can be made to an IP stream, VoIP is possible. Softswitch technologies make managing voice traffic over an IP network possible.

For the application of Vo802.11, it is important to first assess where demand occurs. The largest and most effective opportunities for *Wireless Fidelity* (Wi-Fi) telephony are in the enterprise and vertical markets. In these areas, *wireless LANs* (WLANs) have been adopted and mobile users are easily identified. These businesses have control over the coverage area, bandwidth utilization, and QoS implementation. The users have access to the corporate telephone system, giving Wi-Fi users the same features and accessibility as their wired peers. It is cost effective and efficient to leverage the investment in WLANs by adding wireless telephones. When the cost of Wi-Fi telephones can match what is already available at the local superstore, and low-cost residential IP telephone service is readily available, Wi-Fi telephony will emerge strong in the home. The following sections describe enterprise architectures that could potentially morph from enterprise to residential applications.[15]

Vo802.11 Phones

A more hybrid approach is Vo802.11 technology, such as the wireless PBX interfaces from SpectraLink and Symbol. In this scenario, 802.11 handsets interface with an 802.11 gateway located in front of the legacy PBX. The wireless handsets free employees to roam around the premises (such as medical professionals in a hospital, warehouse workers, or knowledge workers on a corporate campus). Figure 6-9 details that architecture.

According to a recent Cahners-InStat report, additional demand from verticals such as education, healthcare, retail, and logistics will help the overall voice over WLAN market expand to over 80,000

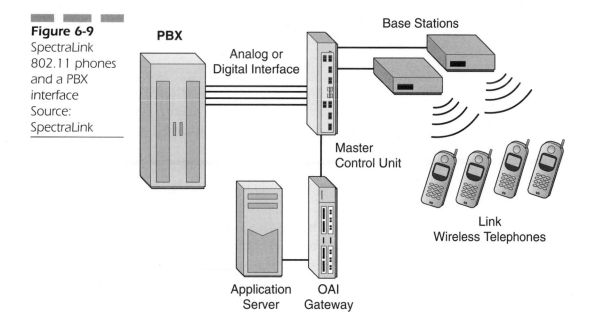

Figure 6-9
SpectraLink
802.11 phones
and a PBX
interface
Source:
SpectraLink

[15]Geri Mitchell, "Wi-Fi Telephony: Wireless Voice over IP Presents New Opportunities, Challenges," a white paper from SpectraLink, www.spectralink.com May 2002.

handset shipments in 2002, a significant jump from the 20,000 shipments in 2001. Furthermore, InStat/MDR reports that annual shipments of Vo802.11x handsets are expected to pass half a million
units by 2006.[16]

At the time of this writing, the retail price of some Vo802.11 handsets is about $700 each. This limits their market to certain verticals
such as healthcare, education, and warehousing where a demon-
strable tradeoff exists between the expense of handset versus the
employee being mobile in the workplace. The limited market for
Vo802.11 handsets keeps prices high. If the technology were to reach
a mass market, driving the production of those handsets, then the
cost per handset would drop.

One potential solution to the high cost of Vo802.11 handsets is to
use VoIP software on wireless modem-equipped PDAs. Although
these high-end PDAs cost upwards of $600, they offer great utility in
being PDAs, telephones, *Short Message Service* (SMS) devices, and
MP3 players.

PDA as an Office Phone via an IP-PBX

A number of ways are available for transmitting voice in 802.11 networks. The simplest approach is to treat the 802.11 network like any
other IP stream and route VoIP using VoIP equipment. Software
products from Pocket Presence and Global IP Sound reside on handheld devices such as Compaq's 802.11-capable iPaq. These devices
present an evolution where voice is simply one more computing function of a PDA.

One barrier to the acceptance of the PDA as a Vo802.11 device has
been incorporating call control. One solution is to incorporate an IP-
PBX using PDAs as Vo802.11 handsets. New-generation IP-PBXs
are more cost competitive (at less than $100 per user), resulting in a
return on investment (ROI) in excess of 100 percent per month.
Equally important, they are easy to install and operate (circuit-
switched PBXs are not).

---

[16]Cahners-InStat, "Voice over Wireless LAN: 802.11x Hears the Call for Wireless
VoIP," April 23, 2002.

These newer IP-PBXs use PDAs as office telephones. Many PDAs contain a sound card with both a microphone and a speaker. Connecting an ear-bud to a Compaq iPAQ PDA, for example, gives it functionality comparable to a cell phone. Software on the PDA converts analog voice to a digital stream that is routed via the IP-PBX to its final destination, which may be another PDA, a computer, or a regular phone. When the voice traffic is routed to another PDA or computer, it travels over the Internet or corporate network, thus saving toll charges.

The PDA can communicate with an IP-PBX using a standard Wi-Fi card, thus making it mobile within the office or wherever a Wi-Fi hot spot is available. The screen of the PDA makes a software-based phone, or softphone, very attractive. The *graphical user interface* (GUI) is much easier to use than a standard phone (or even an IP phone). Some softphones also provide for self-service, where options can be personalized, such as a voice-mail greeting or on-hold music without calling the telephone system administrator (who probably would have to call the vendor or the distributor). Some of the softphone packages also provide advanced communication features such as identifying the participants and speaker(s) in a conference call (see Figure 6-10).[17]

Figure 6-10
A PDA loaded with softphone is a cost-effective Vo802.11 handset.

---

[17]Percy Rajani, "An IP-PBX Helps Turn Your PDA Into an Office Phone," *PDA Planet Magazine*, an article available online at www.planetpdamag.com/content/103002ve.htm October 30, 2002.

Case Study: AmberWaves WISP Vo802.11

AmberWaves is a WISP in northwest Iowa. One of their clients has 3 offices linked by an 802.11 network with 35 employees. The greatest distance between the 3 offices is 19 miles.

This wireless network enables the firm to be its own internal data and voice service provider. The use of Vo802.11 frees the firm from local and long-distance telephone bills. The end users report that the QoS on the network is better than the frame relay circuit they previously used (see Figure 6-11).

Conclusion

At the time of this writing, the *Regional Bell Operating Companies* (RBOCs) of North America are losing thousands of lines per month, the first loss in coverage percentage-wise since the Great Depression. Most of the blame for these losses can be placed on cell phones. SBC Communications reports a loss of 3 million phone lines (called

Figure 6-11
Linking offices
with Vo802.11

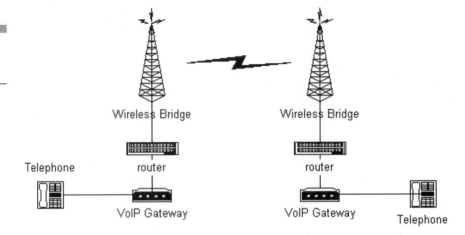

access lines) between 2000 and 2002; they say they will lose another 3 million lines in 2003.[18] Some local phone service providers claim this is due to an unfair regulatory regime where the RBOCs are forced to provide *unbundled network elements* (UNEs) at what they claim are disfavorable rates. Other market analysts point to a number of influences.

Cells phones have claimed a number of those landlines. Many subscribers find cell phones more convenient and have "cut the wire." The monthly subscription cost of a cell phone has dropped and many subscribers are dropping their landline accounts in favor of their cell phone. Questions regarding a comparison of the QoS landline versus a cell phone seem to weigh in favor of the cell phone's convenience over the landline's reputation for reliability and QoS.

Another explanation for one of the leading RBOCs' loss of almost 6 million lines in a little over 2 years is broadband. Given that the traditional 64 Kbps copper pair service provided to the majority of North American residences and small businesses was designed entirely for voice service with a limited data service capability (56 Kbps with 10 percent of households able to receive *Digital Subscriber Line* (DSL) service at data rates around 256 Kbps), there does not appear to be an appreciable level of future-proofing built into the PSTN. Many DSL and cable TV subscribers have either taken up VoIP applications for their voice service or rely on cell phones for voice and their broadband connection for Internet access. In the case of DSL subscribers, they cancel second phone lines that they had with their phone company and use the primary line for both DSL and telephone. Cable subscribers cancel their landline altogether.

The key here is that consumers are constantly questioning the market and are willing to try alternative service providers. Many have already given up a wired phone for a wireless (cell) phone. Why

---

[18]Vincent Vittore and Glenn Bischoff, "Access Line Count Evaporating," *Telephony Magazine* (October 14, 2002): 8.

then would they not give up copper wire access for wireless access delivering voice and high-speed data? In summary, 802.11 presents the best of all possible worlds for the *small office/home office* (SOHO) subscriber in providing telephony as good or better than the PSTN while delivering an overwhelming advantage in bandwidth.

Considerations in Building 802.11 Networks

The successful deployment of an 802.11 system requires design, planning, implementation, operation, and maintenance. This chapter is provided as a very brief overview of what the wireless network planner needs to deploy an 802.11 network.

Design

Many vendors are available with equipment that uses several different 802.11 standards. How does one make sense of it all? The most common protocol in use today is 802.11b. But is it best for my application? Many *wireless Internet service providers* (WISPs) such as T-mobile have phased out all 802.11 (FH) *access points* (APs). Much of this gear is available on the surplus market. Is there any merit to using this technology? Two new protocols are currently emerging: 802.11g and 802.11a. Is there a reason to wait for these to become publicly available and fully certified before interworking them? The short answer to this question is that it depends on the application and the requirements for that application.

Some of the questions that must be addressed in selecting an 802.11 solution lead to trade-offs, such as speed versus range. Others have mutually exclusive answers, such as proprietary- versus standards-based extensions. Some of the questions that need to be addressed are: What is the network topology? What kinds of links will be used? What is the environment like? What are the throughput, range, and bit error rate that are needed? Will you need tolerance for multipath? Which frequency band will be used with which protocols? Can the solution be off-the-shelf or surplus standards-based, or will it need to be custom?

Network Topology

One of the goals of network planning is to assure that work gets done within the budget allotted to the project. All nodes should be able to communicate where they need to at any time. Network designs can be redundant at various levels so that if a node fails, no other node should be affected. If security is a concern, trusted parts of the network must be separated from the untrusted parts. One of the major factors that determines throughput, robustness, reliability, security, and cost is the geometric arrangement of the network components, or the topology.

Five major topologies are in use today in wired networks: bus, star, tree, ring, and mesh. In a *wireless local area network* (WLAN), only the star and mesh have analogues with the wired networks. These topologies can be implemented using the modes of operation supported by the *Institute of Electrical and Electronics Engineers* (IEEE) 802.11 standards, infrastructure and ad hoc, through their service sets the *independent basic service set* (IBSS), the *basic service set* (BSS), and the *extended service set* (ESS). See Figure 7-1.

Currently, the most common mode is infrastructure. In this mode, wireless devices can communicate with each other or with a wired network. When an AP is connected to wired network and a set of wireless stations, it is referred to as a BSS. A BSS consists of at least one AP connected to the wired network infrastructure and a set of wireless end stations. Thus, BSS configurations rely on an AP that acts as a switch for a single WLAN cell or channel. An ESS is a set of two or more BSSs, each containing an AP connected together using a *distribution system* (DS) to form a single subnet. Although the DS could be any type of network, it is often an Ethernet LAN. A mobile user can move between APs and reassociate with the AP providing

Figure 7-1
Service sets
available in
802.11
standards

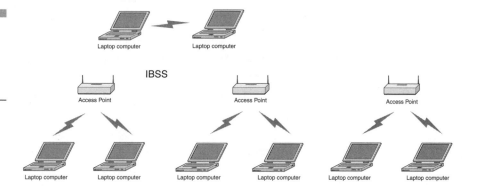

the best coverage. In this way, seamless coverage is possible within the subnet. Most WLANs operate in infrastructure mode because they require access to the wired LAN for services like file servers, printers, and Internet access.

In ad hoc mode, devices or stations communicate directly with each other, without the use of an AP. Ad hoc mode is also referred to as peer-to-peer mode and uses the IBSS. IBSS configurations are also referred to as an independent configuration or an ad hoc network. Logically, an IBSS configuration is analogous to a peer-to-peer office network in which no single node is required to function as a switch or router. IBSS WLANs include a number of nodes or wireless stations that communicate directly with one another on an ad hoc, peer-to-peer basis. Thus, it contains a set of wireless stations that communicate directly with one another without using an AP or any connection to a wired network. It is useful for quickly and easily setting up a wireless network anywhere a wireless infrastructure does not exist or is not required for services. This could include a hotel room, convention center, or airport, or where access to the wired network is barred (such as for consultants at a client site). Generally, IBSS implementations cover a limited area and aren't connected to any larger network. An example would be two laptops with 802.11b *Personal Computer Memory Card International Association* (PCM-CIA) cards that want to share files.

The star topology, which happens to be in widest use today, is a network in which one central base station or AP is used for communication. Information packets are transmitted by the originating node and are received and routed by the AP to the proper wireless destination node by the AP. See Figure 7-2.

The mesh topology is a slightly different type of network architecture than the star topology, except that no centralized base station exists. Each node that is in range of another one can communicate freely, as shown in Figure 7-3.

Wireless mesh networks are an exciting new topology for creating low-cost, high-reliability wireless networks indoors, across a campus, or in a metropolitan area. In a mesh network, each wireless node serves as both an AP and a wireless router, creating multiple pathways for the wireless signal. Mesh networks have no single point of failure and thus are self-healing. A mesh network can be designed to route around line-of-sight obstacles that can interfere with other

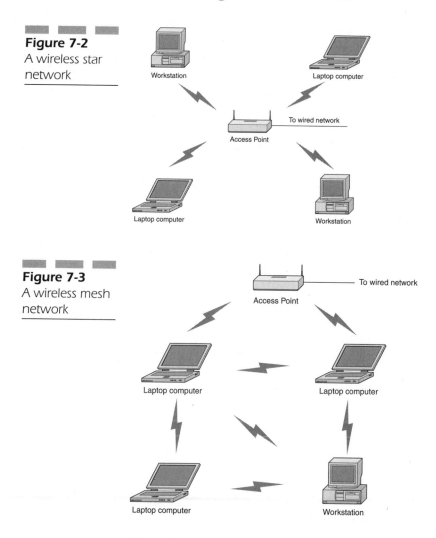

Figure 7-2
A wireless star network

Figure 7-3
A wireless mesh network

wireless network topologies. However, a wireless mesh currently requires the use of specialized client software that will provide the routing function and put the radio into ad hoc or infrastructure mode as required.

Link Type

802.11 systems can be built using either point-to-point or point-to-multipoint links. The FCC regulations allow both types of links, but they come with power-to-the-antenna implications.

Environment

What is the environment like? Is it indoors, or is it outdoors? Is there a line of sight or are there obstacles in the path?

Throughput, Range, and Bit Error Rate

Throughput has tradeoffs with range and the *bit error rate* (BER). The best network designs balance these factors by limiting the data rate according to data quantity and latency requirements. Fundamentally, any application will trade off between three factors: range, throughput, and BER. 802.11 was developed as a wireless replacement for an indoor Ethernet. If the primary indoor application is Internet access with the AP less than 100 feet away, then any one of the 802.11 standards will work. Since the throughput is limited by the protocol, and BER has to be reasonably high to get throughput, the only variable left is the range. The available range at a given throughput can be calculated using a link budget.

Multipath Fading Tolerance

Non line of sight communications in the *Industrial, Scientific, and Medical* (ISM) and *Unlicensed National Information Infrastructure* (U-NII) bands must allow for significant multipath fading. Multipath is created by reflections canceling the main signal. The choice of frequency band and protocols will in part depend on how much multipath can be tolerated.

Link Budget

A fundamental concept in any communications system is the link budget. The link budget is a summation of all the gains and losses in a communications system. The result of the link budget is the transmit power required to present a signal with a given *signal-to-noise ratio* (SNR) at the receiver to achieve a target BER.

For *wireless fidelity* (Wi-Fi), it is sufficient to consider factors such as path loss, noise, receiver sensitivity, and gains and losses from antennas and cable. Before calculating a link budget, factors like the frequency band must be determined.

Frequency Band

802.11 technologies can be deployed on four unlicensed frequency bands in the two ISM and U-NII bands. The 2.4 GHz ISM band has an inherently stronger signal with a longer range and can travel through walls better than the 5 GHz U-NII bands. However, the U-NII band enables more users to be on the same channel simultaneously. The 2.4 GHz ISM band has a maximum of three nonoverlapping 22 MHz channels, while the 5 GHz band has four nonoverlapping 20 MHz channels in each of the U-NII bands.

ISM Band   The ISM bands were originally reserved internationally for the noncommercial use of *radio frequency* (RF) electromagnetic fields for industrial, scientific, and medical purposes. More recently, they have also been used for license-free, error-tolerant communications applications such as cordless phones, Bluetooth, and Wireless WLAN.

U-NII Band   The U-NII bands can be used by devices that will provide short-range, high-speed wireless digital communications. These devices, which do not require licensing, will support the creation of WLANs and facilitate access to the Internet. The U-NII spectrum is located at 5.15 to 5.35 GHz and 5.725 to 5.825 GHz.

The 5.15 to 5.25 GHz portion of the U-NII band is intended for indoor, short-range networking devices. The FCC adopted a 200 *milliwatt* (mW) *effective isotropic radiated power* (EIRP) limit to enable short-range WLAN applications in this band without causing interference to *mobile satellite service* (MSS) feeder link operations.

Devices operating between 5.25 to 5.35 GHz are intended to provide communications within and between buildings, such as campustype networks. U-NII devices in the 5.25 to 5.35 GHz frequency range are subject to a 1-watt EIRP power limit.

The 5.725 to 5.825 GHz portion of the U-NII band is intended for community networking communications devices operating over longer distances. The FCC permits fixed, point-to-point U-NII devices to operate with up to a 200-watt EIRP limit.

FCC Regulations The use of these bands is regulated under Parts 15.247 and 15.407 of the FCC regulations.[1] The following are the relevant passages of Part 15.247 regarding power at the time of writing:

(b) The maximum peak output power of the intentional radiator shall not exceed the following:

(1) For frequency hopping systems operating in the 2400–2483.5 MHz or 5725–5850 MHz band and for all direct sequence systems: 1 watt.

(3) Except as shown in paragraphs (b)(3) (i), (ii) and (iii) of this section, if transmitting antennas of directional gain greater than 6 dBi are used the peak output power from the intentional radiator shall be reduced below the stated values in paragraphs (b)(1) or (b)(2) of this section, as appropriate, by the amount in dB that the directional gain of the antenna exceeds 6 dBi.

(i) Systems operating in the 2400–2483.5 MHz band that are used exclusively for fixed, point-to-point operations may employ transmitting antennas with directional gain greater than 6 dBi provided the maximum peak output power of the intentional radiator is reduced by 1 dB for every 3 dB that the directional gain of the antenna exceeds 6 dBi.

(ii) Systems operating in the 5725–5850 MHz band that are used exclusively for fixed, point-to-point operations may employ transmitting antennas with directional gain greater than 6 dBi without any corresponding reduction in transmitter peak output power.

Part 15.407 regulates the UNII band and its operation. The following parts are the relevant passages from Part 15.407 for understanding the power limits within the 5.1, 5.2, and 5.8 GHz bands:

---

[1]The FCC web site, www.fcc.gov, has a lot of material online. Part 15 in its entirety can be found at www.access.gpo.gov/nara/cfr/waisidx_01/47cfr15_01.html.

(a) Power limits:

(1) For the band 5.15–5.25 GHz, the peak transmit power over the frequency band of operation shall not exceed the lesser of 50 mW or 4 dBm + 10logB, where B is the 26-dB emission bandwidth in MHz. In addition, the peak power spectral density shall not exceed 4 dBm in any 1-MHz band. If transmitting antennas of directional gain greater than 6 dBi are used, both the peak transmit power and the peak power spectral density shall be reduced by the amount in dB that the directional gain of the antenna exceeds 6 dBi.

(2) For the band 5.25–5.35 GHz, the peak transmit power over the frequency band of operation shall not exceed the lesser of 250 mW or 11 dBm + 10logB, where B is the 26-dB emission bandwidth in MHz. In addition, the peak power spectral density shall not exceed 11 dBm in any 1-MHz band. If transmitting antennas of directional gain greater than 6 dBi are used, both the peak transmit power and the peak power spectral density shall be reduced by the amount in dB that the directional gain of the antenna exceeds 6 dBi.

(3) For the band 5.725–5.825 GHz, the peak transmit power over the frequency band of operation shall not exceed the lesser of 1 W or 17 dBm + 10logB, where B is the 26-dB emission bandwidth in MHz. In addition, the peak power spectral density shall not exceed 17 dBm in any 1-MHz band. If transmitting antennas of directional gain greater than 6 dBi are used, both the peak transmit power and the peak power spectral density shall be reduced by the amount in dB that the directional gain of the antenna exceeds 6 dBi. However, fixed point-to-point U-NII devices operating in this band may employ transmitting antennas with directional gain up to 23 dBi without any corresponding reduction in the transmitter peak output power or peak power spectral density. For fixed, point-to-point U-NII transmitters that employ a directional antenna gain greater than 23 dBi, a 1 dB reduction in peak transmitter power and peak power spectral density for each 1 dB of antenna gain in excess of 23 dBi would be required. Fixed, point-to-point operations exclude the use of point-to-multipoint systems, omni directional applications, and multiple collocated transmitters

transmitting the same information. The operator of the U-NII device, or if the equipment is professionally installed, the installer, is responsible for ensuring that systems employing high-gain directional antennas are used exclusively for fixed, point-to-point operations.

Table 7-1 summarizes the ISM and U-NII unlicensed frequency bands used by Wi-Fi devices. The table also shows their associated power limits.

Point to Multipoint   Part 15.247(b)(1) limits the maximum power at the antenna to 1 watt. Part 15.247(b)(3) allows antennas that have more than 6 *decibels* (db) as long as the power to the antenna is reduced by an equal amount in the 2.4 GHz band. This implies that the maximum EIRP is 4 watts or 36 dBm. This limit of 4 watts EIRP no matter the antenna gain is illustrated in Table 7-2.

Point-to-Point Links   Point-to-point links have a single transmitting point and a single receiving point. Typically, a point-to-point link is used in a building-to-building application. Part 15.247 (b)(3)(i) allows the EIRP to increase beyond the 4-watt limit for point-to-multipoint links in the 2.4 GHz ISM band. For every additional 3 db

Table 7-1

Frequency bands and associated power limits.

Frequency range (MHz)	Bandwidth (MHz)	Max. power at antenna	Max. EIRP	Notes
2400–2483.5	83.5	1 W (+30 *dB above 1 milliwatt* [dBm]), 1 W (+30 dBm)	4 W (+36 dBm)	Point to point; point to multi-point following 3:1 rule
5150–5250	100	50 mW	200 mW (+23 dBm)	Indoor use, must have integral antenna
5250–5350	100	250 mW (+24 dBm)	1 W (+30 dBm)	
5725–5825	100	1 W (+30 dBm)	200 W (+53 dBm)	

Table 7-2

Point-to-multipoint operation in the 2.4 GHz ISM band

Power at antenna (mW)	Power at antenna (dBm)	Max. antenna gain (dBi)	EIRP (watts)	EIRP (dBm)
1000	30	6	4	36
500	27	9	4	36
250	24	12	4	36
125	21	16	4	36
63	18	19	4	36
31	15	21	4	36
15	12	24	4	36
8	9	27	4	36
4	6	30	4	36

gain on the antenna, the transmitter only needs to be cut back by 1 dB. The so-called three-for-one rule for point-to-point links can be observed in Table 7-3.

According to part 15.247(b)(3) (ii), the 5.8 GHz band has no such restriction. However, part 15.407 effectively restricts the EIRP to 53 dBm (see Table 7-4).

Protocols

Four primary standards-based protocols are currently available: 802.11 802.11b, 802.11a, and 802.11g.

802.11 The 802.11 standard was the first standard to specify the operation of a WLAN. This standard addresses *frequency-hopping spread spectrum* (FHSS), *direct sequence spread sequence* (DSSS), and infrared. The data rate is limited to 2 Mbps and 1 Mbps for both FHSS and DSSS.

FHSS handles multipath and narrowband interference as well as a byproduct of its frequency-hopping scheme. If multipath fades one

Table 7-3

Point-to-point operation in the 2.4 GHz ISM band

Power at antenna (mW)	Power at antenna (dBm)	Max. antenna gain (dBi)	EIRP (watts)	EIRP (dBm)
1000	30	6	4	36
794	29	9	6.3	38
631	28	12	10	40
500	27	15	16	42
398	26	18	25	44
316	25	21	39.8	46
250	24	24	63.1	48
200	23	27	100	50
157	22	30	157	52

Table 7-4

Point-to-point operation in 5.8 GHz U-NII band

Power at antenna (mW)	Power at antenna (dBm)	Antenna gain (dBi)	EIRP (watts)	EIRP (dBm)
1000	30	6	4	36
1000	30	9	8	39
1000	30	12	16	42
1000	30	15	316	45
1000	30	18	63.1	48
1000	30	21	125	51
1000	30	23	250	53

channel, other channels are usually not faded. Thus, packets are passed on those hops where no fading occurs. Operating an FHSS system in a high-multipath or high-noise environment will be seen as an increase in latency. FHSS has 64 hopping patterns that can support up to 15 colocated networks. FHSS systems are limited to 1 Mbps and optionally 2 Mbps. Typically, they have a shorter range than DSSS systems. FHSS is not compatible with today's 802.11b equipment.

DSSS, as implemented in 802.11, occupies 22 MHz of spectrum while providing a maximum over-the-air data rate of 2 Mbps. DSSS is susceptible to multipath and narrowband interference due to the limited amount of spreading that is used (11 bits). DSSS can only support three noninterfering channels and thus does not have nearly as much network capacity as an FHSS system at the same data rate. DSSS is compatible with today's 802.11b equipment.

Surplus 802.11 equipment may work well for some applications where multipath immunity is required, lower data rates can be tolerated, and compatibility with currently available equipment is not desired. Furthermore, be advised that the gear may no longer be covered by warranties and may not have service available for it anymore.

802.11b   The most widely used standard protocol, 802.11b, requires DSSS technology, specifying a maximum over-the-air data rate of 11 Mbps and a scheme to reduce the data rate when higher data rates cannot be sustained. This protocol supports 5.5 Mbps, 2 Mbps, and 1 Mbps over-the-air data rates in addition to 11 Mbps using DSSS and *complementary code keying* (CCK).

IEEE 802.11b standard uses CCK as the modulation scheme to achieve data rates of 5 and 11 Mbps. 802.11 reduced the spreading from 11 bits down to 8 to achieve the higher data rates. The modulation scheme makes up the processing gain lost with the lower spreading by using more forward error correction.

The IEEE 802.11b specification allows for the wireless transmission of approximately 11 Mbps of raw data at indoor distances up to 300 feet and outdoor distances of 20 miles in point-to-point usage of the 2.4 GHz band. The distance depends on impediments, materials, and line of sight.

802.11b is the most commonly deployed standard in public short-range networks, such as those found at airports, coffee shops, hotels, conference centers, restaurants, bookstores, and other locations. Several carriers currently offer pay as you go hourly, per session, or with unlimited monthly access using networks in many locations around the United States and other countries.

802.11a   The 802.11a standard operates in the three 5 GHz U-NII bands and thus is not compatible with 802.11b. The bands are designated by application. The 5.1 GHz band is specified for indoor use only, the 5.2 GHz band is designated for indoor/outdoor use, and the 5.7 Ghz band is designated for outdoors only. RF interference is much less likely because of the less crowded 5 GHz bands. The 5 GHz bands each have four separate nonoverlapping channels. They specify *orthogonal frequency division multiplexing* (OFDM) using 52 subcarriers for interference and multipath avoidance, they support a maximum data rate of 54 Mbps using 64 *quadrature amplitude modulation* (QAM), and they mandate support of 6, 12, and 24 Mbps data rates. The protocol specifies the minimum receive sensitivities ranging from −65 dBm for the 54 Mbps rate to −82 dBm for 6 Mb/sec. Equipment designed for the 5.1 GHz band has an integrated antenna and is not easily modified for higher power output and operation on the other two 5 GHz bands.

802.11g   802.11g is an extension to 802.11b and operates in the 2.4 GHz band. 802.11g increases 802.11b's data rates to 54 Mbps using the same OFDM technology used in 802.11a. The range at 54 Mbps is less than the existing 802.11b APs operating at 11 Mbps. As a result, if an 802.11b cell is upgraded to 802.11b, the high data rates will not be available throughout all areas. You'll probably need to add additional APs and replan the RF frequencies to split the existing cells into smaller ones. 802.11g offers higher data rates and more multipath tolerance. Although more interference exists on the 2.4 GHz band, 802.11g may be the protocol of choice for the best range and bandwidth combination. It's also upwardly compatible with 802.11b equipment.

Making the Choice

Which technique is best? It depends on the application and other design considerations. Frequency hopping offers superior reliability in noise and multipath fading environments. Direct sequence can provide higher over-the-air data rates. OFDM offers multipath tolerance and much higher data rates. 802.11 (no letter) is now obsolete, but may offer nonstandard, bargain-basement usable equipment. 802.11b is compatible with most of the public access locations. 802.11a is the best to solve interference cases and has great throughput. 802.11g promises the best range and throughput combination of all the solutions.

Standard or Custom?

Interoperability with other manufacturers' products requires a standard protocol. However, some manufacturers have products that offer features which others don't. For example, Panasonic sells a residential gateway and a client card that use WhiteCap, a technology sold by ShareWave, now part of Cirrus Logic. WhiteCap is an intermediate solution that eventually may upgrade to IEEE 802.11e. To take advantage of the *quality of service* (QoS) provided by WhiteCap, both the APs and the client adapter cards have to support the extra technology. The chance exists that it may not be possible to buy additional components if the company stops producing the product.

Planning

Understanding what needs to be covered is an important part of the planning process, so do some quick calculations to better understand the situation. For indoor coverage, understand the effects of the attenuating and reflecting materials where the coverage is desired. A link budget will tell you what is practical given the environment and how to plan cells. With a link budget, one can have an estimate of how many cells will be required for the project. Tradeoffs take place between more cells and running more power. For an outdoor application, also consider checking the Fresnel zone. For example, the tradeoff between working on one long-distance shot versus two back-to-back links can be discovered by working out a few things on paper first.

Fresnel Zone   For a point-to-point shot, it is important to understand the effects of the Fresnel Zone. For an indoor application, this can be skipped since the losses are calculated into the link budget. So what is the Fresnel Zone?

In a Fresnel Zone, the signal extends from the transmitting antenna to form an ever-expanding cone. Although some of the signal travels directly in the line of sight along the center of the cone, other parts of the signal reradiate off points along the way. At the receiver, signals from the direct line of sight indirectly cancel and add to each other. The first Fresnel Zone is the surface containing every point for which the sum of the distances from those points to the two ends of the path is exactly a half wavelength longer than the direct end-to-end path. The second Fresnel Zone is the surface containing every point for which the sum of the distances from those points to the two ends of the path is exactly one wavelength longer than the direct end-to-end path. Figure 7-4 shows the first and second Fresnel Zones.

One can calculate the perpendicular distance of the Fresnel Zone from the line that connects the transmitter and the receiver using the following n^{th} Fresnel Zone formula:

Figure 7-4

First and second Fresnel Zones

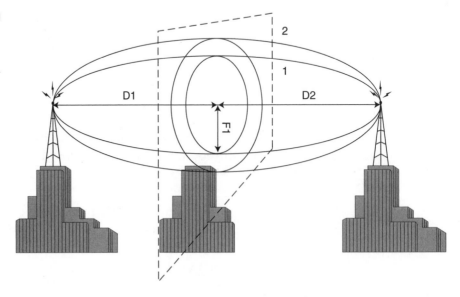

$$F_N = \sqrt{\frac{N \times \lambda \times D_1 \times D_2}{D_1 + D_2}} \qquad (7\text{-}1)$$

where

N is the Fresnel Zone number; $N = 1$ is the first Fresnel Zone.

λ is the wavelength (meters).

D_1 and D_2 are the distances to the endpoints (meters).

If reflections of the signal are made from an odd-numbered Fresnel Zone, the signal level will cancel at the receiver, but if the reflection is from an even-numbered Fresnel Zone, it will add at the receiver. Therefore, on long-distance shots, it is necessary to take into account ground/water reflections and vertical surfaces like tall buildings.

Since the majority of the transmitted power is in the first Fresnel Zone, any time the path clearance between the terrain and the line-of-sight path is less than $0.6F_1$ (six-tenths of the first Fresnel Zone distance), some knife-edge diffraction loss will occur. The amount of loss depends on the amount of penetration. To find out if any obstruction is taking place in the Fresnel Zone, a profile of the terrain is superimposed with the ellipse created by the nth Fresnel Zone formula with the first Fresnel Zone ($N=1$) and multiplying the result by 0.6 for repetitive points across the profile.

It is possible to gain in the signal strength at the receiver up to 3 dB by having a flat surface such as a lake, a highway, or a smooth desert area at the second Fresnel Zone in such a way that the signals get reinforced at the receiver.

Decibels and Signal Strength Rather than keep track of all those zeroes, amplifier power is measured in a logarithmic scale using dB. Then, instead of multiplying and dividing all the gains, it's much simpler to add and subtract dBs.

$$DB = 10 \times \log10 \ (\text{power out/power in}) \qquad (7\text{-}2)$$

Decibel readings are positive when the output is larger than the input, and they are negative when the output is smaller than the input. Each 10 dB change corresponds to a factor of 10, and 3 dB changes are a factor of 2. Thus, a 33 dB change corresponds to a factor of 2000:

$$33 \text{ dB} = 10 \text{ dB} + 10 \text{ dB} + 10 \text{ dB} + 3 \text{ dB} = 10 \times 10 \times 10 \times 2 = 2000$$

Power is sometimes measured in dBm. To find the dBm ratio, simply use 1 mW as the input in the first equation. It's helpful to remember that doubling the power is a 3 dB increase. A 1 dB increase is roughly equivalent to a power increase of 1.25, and a 10 dB increase in power is a power increase of 10 times. With these numbers in mind, quickly perform most gain calculations in your head.

How to Calculate a Link Budget

Link budget planning is an essential part of the network-planning process for both indoor and outdoor applications. A link budget helps to dimension the required coverage, capacity, and QoS requirement in the network. In a typical 802.11 link, two link budget calculations must be made: the link from the AP to the client adapter card, and the other link from the client adapter card back to the AP. Link budgets can be used to determine if the link design meets the designer's criteria for range, throughput, and BER.

A link budget basically adds all the gains and losses to the transmitter power (in dB) to yield the received power. In order to have adequate signal at the receiver, the power presented to the receiver must be at least as much as the receive sensitivity. The link budget for the AP to the client adapter card is shown in Figure 7-5.

Figure 7-5
Calculating a link budget

Access Point   Access Point Antenna   Network Interface Card Antenna   Laptop computer

Transmiter Power and Antenna Gain The combination of power output at the antenna and the gain of the antenna itself are legally limited by FCC rules and regulations in the U.S. In other countries, similar regulations exist but will differ. Page 206 Chapter 7 contains an explanation of the legal limits.

For Wi-Fi, the maximum power output and antenna gain is limited based on what frequency band is used and whether the application is point to point or point to multipoint. The manufacturer of an access point or client adapter card will typically specify the output power in their spec sheet in milliwatts and dBm—decibels over a one milliwatt reference.

To convert from milliwatts to dBm:

$$dBM = log \text{ (power in milliwatts / 1 milliwatt)}$$

So for example, how many dBm is 4 watts? 4 watts = 4000 milliwatts.

$$log \text{ (4000 mW / 1 mW)} = log\ 4000 = 36\ dBm$$

Three good rules of thumb to convert into dB are:

1) A doubling is 3dB

2) 10 times is 10 dB

3) Decibels add where factors multiply

So log of 4000 can be calculated as follows:

$$2 \times 2 \times 10 \times 10 \times 10 = 3\ dB + 3\ dB + 10\ dB$$
$$+ 10\ dB + 10\ dB \text{ or } 36\ dB$$

The m in dBm stands for the reference of one milliwatt.

For antennas, the gains is usually specified in dBi, or decibels over isotropic. The i in dBi stands for the reference of an isotropic antenna.

So for example we can calculate the effective radiated power of a 23 dB gain antenna being fed with 100 milliwatt or 20 dBm source (client adapter card or access point)

$$20\ dBm + 23\ dB = 46\ dBm$$

$$46\ dBm = 2 \times 2 \times 10 \times 10 \times 10 \times 10 \text{ or } 40,000 \text{ milliwatts or } 40\ W$$

Path Loss   The most difficult part of calculating a link budget is the path loss. Outdoors, the free-space loss is well understood. The path loss equation[2] for outdoors can be expressed as

$$\text{Free space path loss} = 20 \log (d \text{ [meters]}) + 20 \log (f \text{ [MHz]})$$
$$+ 36.6 \text{ dB} \tag{7-3}$$

At 2.4 GHz, the formula simplifies to

$$\text{Free space path loss} = 20 \log (d \text{ [meters]}) + 40 \, dB \tag{7-4}$$

This formula holds true as long as one can see along the line of sight from the receiver and the transmitter, and one has a sufficient amount of area around the Fresnel Zone. Indoors, this formula is more complicated and depends on factors such as building materials, furniture, and occupants.

At 2.4 GHz, one estimate follows this formula:

$$\text{Indoor path loss } (2.4 \, GHz) = 55 \text{ dB} + 0.3 \text{ dB/d [meters]} \tag{7-5}$$

At 5.7 GHz, the formula looks like this:

$$\text{Indoor path loss } (5.7 \text{ GHz}) = 63 \text{ dB} + 0.3 \text{ dB/d [meters]} \tag{7-6}$$

Receive Antenna Gain   The receive antenna gain adds into the link budget just like the transmit antenna. Adding gain to an antenna is balanced gain since it adds gain for both transmit and receive.

Link Margin   Fade margin is the difference, in dB, between the magnitude of the received signal at the receiver input and the minimum level of signal determined for reliable operation. The higher the fade margin, the more reliable the link will be. The exact amount of fade margin required depends on the desired reliability of the link,

---

[2]Edward C. Jordan, ed. *Reference Data for Engineers: Radio, Electronics, Computer, and Communications* (Howard W. Sams and Co., Indianapolis, IN 1986).

but a good rule-of-thumb is 20 to 30 dB. Fade margin is often referred to as the thermal or system operating margin.

Diffraction Losses Diffraction occurs when the radio path between the transmitter and the receiver is obstructed by a surface that has sharp irregularities or an edge. The secondary waves resulting from the obstructing surface are present behind the obstacle. When close to line of sight, diffraction losses can be as little as 6 dB. In non line of sight obstacles, diffraction losses can be 20 to 40 dB.

Coax and Connector Losses Connector losses can be estimated at 0.5 dB per connection. Cable losses are a function of cable type, thickness, and length. Generally speaking, the thicker and better built the cable, the lower are the losses (and the higher the cost). As can be seen from Table 7-5, coax losses are nearly prohibitive in the 2.4 and 5.8 GHz bands. The best option is to use coax as sparingly

Table 7-5

Coax attenuation losses (Source: Andrew, Times Microwave and Belden online catalogs)*

Cable type	2.4 GHz		5.8 GHz	
	dB/100 ft	dB/100 m	dB/100 ft	dB/100m
RG-58	32.2	105.6	51.6	169.2
RG-8X	23.1	75.8	40.9	134.2
LMR-240	12.9	42.3	20.4	66.9
RG213/214	15.2	49.9	28.6	93.8
9913	7.7	25.3	13.8	45.3
LMR-400	6.8	22.3	10.8	35.4
³/₈″ LDF	5.9	19.4	8.1	26.6
LMR-600	4.4	14.4	7.3	23.9
¹/₂″ LDF	3.9	12.8	6.6	21.6
⁷/₈″ LDF	2.3	7.5	3.8	12.5
1¹/₄″ LDF	1.7	5.6	2.8	9.2
1⁵/₈″ LDF	1.4	4.6	2.5	

*Andrew http://www.andrew.com/products/trans_line/default.aspx
Times Microwave http://www.timesmicrowave.com/telecom/pdf/LMRGuide.pdf
Belden http://bwcecom.belden.com/

as possible and locate the microwave transceiver as close to the antenna as possible in an environmental enclosure.

Attenuation There are a number of atmospheric factors that will diffuse or otherwise disrupt the flow of radio waves through the atmosphere. Those factors include rain and fog, trees, fiberglass, glass and other building materials.

Rain and Fog When deploying in a rainy or foggy climate, it may be necessary to plan additional signal loss due to rain or fog. 2.4 GHz signals may be attenuated by up to 0.05 dB/km (0.08 dB/mile) by heavy rain (4 inches/hr). Thick fog produces up to 0.02 dB/km (0.03 dB/mile) attenuation. At 5.8 GHz, heavy rain may produce up to 0.5 dB/km (0.8 dB/mile) attenuation, and thick fog up to 0.07 dB/km (0.11 dB/mile). Even though rain itself does not cause major propagation problems, rain will collect on the leaves of trees and will produce attenuation until it evaporates.

Trees Trees can be a significant source of path loss,[3] and a number of variables are involved, such as the specific type of tree, whether it is wet or dry, and, in the case of deciduous trees, whether the leaves are present or not. Isolated trees are not usually a major problem, but a dense forest is another story. The attenuation depends on the distance the signal must penetrate through the forest, and it increases with frequency. The attenuation is of the order of 0.35 dB/m at 2.4. This adds up to a lot of path loss if your signal must penetrate several hundred meters of forest.

Fiberglass The loss for a radome, which is a plastic shell that protects an antenna from the elements in outdoor deployments, is about .5 to 1 dB.

---

[3]Anil Shukla, "A Generic Vegetation attenuation model for 1-60GHz: PM3035," www.radio.gov.uk/topics/research/topics/propagation/vegetation/veg-attenuation-model.pdf.

Glass A normal, clear glass pane will lose about 3 dB at 2.4 GHz. Although most glass will not affect radio frequency, certain kinds of glass severely attenuate signals.[4] It depends on the glass and the tint material. If the glass is reflective at all on either side, chances are that a signal may not be able to penetrate it. New construction often uses tinted, coated, or High-E glass that is designed to hold heat out. However, it attenuates 802.11 signals. High-E is energy efficient; is usually double-paned, coated, and filled with argon or other inert gasses; and is not necessarily tinted. Tin oxide (SnO_2) coatings do not pass RF. Some windows have as much as a 20 dB loss.

Other Building Materials Examples of attenuation through various building materials are shown in Table 7-6.[5]

Examples In order to better understand how attenuation degrades wireless applications, examples of degrading instances are offered.

Table 7-6 Attenuation of various building materials		
Window in brick wall	2 dB
Metal frame glass wall into building	6 dB
Office wall	6 dB
Metal door in office wall	6 dB
Cinder block wall	4 dB
Metal door in brick wall	12.4 dB
Brick wall next to metal door	3 dB

[4]"Glass That Cuts Signals," www.isp-planet.com/fixed_wireless/technology/2001/tint_bol.html.

[5]John C. Stein, "Indoor Radio WLAN Performance Part II: Range Performance in a Dense Office Environment," Intersil Corporation, 1997, www.intersil.com/design/prism/papers/symposum.pdf.

Example 1 On the network side is a Cisco 350 AP that operates at its maximum output power at 2.45 GHz into a 6 dB gain omni-directional antenna. On the other side of a 1 km link is a laptop with a Cisco Aironet 350 WLAN adapter card. Is there sufficient signal to make this work?

$\quad$ 20 dBm 100 mW

+ 6 dBi (antenna gain)

− 100 dBm (path loss for 1 km at 2.4 GHz is 40 + 20 log[1000m])

+ 2.2 dBi (antenna gain of client adapter card)

− 20 dB (link margin) $\hfill$ (7-7)

= −91.8 dBm (minimum received power)

Example 2 A company claims that a distance of 4.3 miles or 7 km can be spanned in a point-to-multipoint application with their antenna using 802.11b. Assume no coax losses, no connector losses, and a perfect line of sight. Does it work with a minimum spec card? Does it work with the best card?

$\quad$ 36 dBW 4 W EIRP (maximum power out point to multipoint; includes power out and antenna gain)

− 116.9 dB (path loss for 7 km is 40 + 20 log[7000 m]) $\hfill$ (7-8)

+ 2.2 dBi (antenna gain of client adapter card)

− 20 dB (link margin)

= 98.7 dBm (minimum received power)

The worst-case receive sensitivity for an 802.11b card is −80 dBm at 1 Mbps. Quite clearly, the signal is not adequate, and this is only barely enough to leave a link margin of 1.3 dB, which is not a very reliable link. The Cisco Aironet 250 client adapter card is the most sensitive in the industry (1 Mbps: −94 dBm, 2 Mbps: −91 dBm, 5.5 Mbps: −89 dBm, 11 Mbps: −85 dBm) and will yield a link margin of 15.3 dB using a 1 Mbps link.

But can you get back at the AP?

20 dBm (max. output of Cisco 350 client adapter card)

+ 2.2 dBi (antenna gain of client adapter card)

− 116.8 dBm (path loss for 7 km at 2.4 GHz is 40 + 20 log[7000 m])   (7-9)

− 20 dB (link margin)

+ 18 dB gain (best guess at the gain of a 2 foot × 2 foot phased antenna)

= 96.6 − dBm (minimum received power)

If the link margin is 15 dB, or the antenna has 23 dB gain, the Cisco 350 client adapter card can be heard at 7 km with this antenna.

Given current technology, what is the best you can do on a point-to-point link?

- Get more transmit power and better receive sensitivity.

- Remove noise.

- Limit any attenuation from the link budget.

- Get antennas with the most gain for both ends.

- Go to a lower data rate for better sensitivity (a higher data rate equals less power efficiency).

- Use two antennas for diversity at both ends.

Premium cards like the Cisco Aeronet 350 card have output power at 100 mW or 20 dBm, and a very sensitive receiver at −85 dBm at 11 Mbps.

If you now go to a lower data rate, such as 1 Mbps, the receive sensitivity is at −94 dBm. A link budget for a record-breaking point-to-point link looks like this:

24 dBm (maximum legal output of transmitter)

+ 24 dBi (grid antenna gain)

− 20 dB (link margin)

− 2 dB (connector losses)

− 2 dB (coax losses)

+ 24 dB gain (grid antenna gain)

+ 94 dBm (minimum received power at 1 Mbps)

= maximum path loss = 142 dbm

or about 75 miles (path loss for 120 km at 2.4 GHz
is $40 + 20 \log [120{,}000 \text{ m}]$)

Under full multipath conditions, this link will have a 1 Mb data rate. Under better conditions, the link may operate at the full data rate of 11 Mbps.

Site Plan Prediction Tools

Indoor propagation tools are available, such as WinProp,[6] campus-wide propagation tools like SitePlanner,[7] and ray-tracing tools like CINDOOR[8] that can be used to model buildings and campuses given a *computer-aided design* (CAD) floor plan. One of the shortcomings is that the floor plan does not tell you much about the construction materials used in the structure. For example, is the wall made of bricks, Sheetrock, or reinforced concrete? Is the floor totally RF isolated from the one above or below it? Only a walkthrough will tell you these things accurately.

Site Survey

Once things work on paper with an adequate link budget and the Fresnel Zone, one can go out to the site and see if the paper plan works.

Outdoor Site Survey   All the data on paper may indicate that everything will work for a particular link. The Fresnel Zone can be checked against a topographical view of the point-to-point shot. You

---

[6]AWE Communications, www.awe-communications.com.

[7]Wireless Valley, www.wvcomm.com.

[8]University of Cantebria, Spain, www.gsr.unican.es.

may even use expensive ray-tracing programs to predict the path, but only one way will tell you if the installation will work.

To perform an outdoor site survey for a point-to-point shot, take along binoculars, two-way radios or cell phones, topology maps, a *Global Positioning System* (GPS), a spectrum analyzer, a 5 and 10 dB attenuator, and your radio equipment to take the trial shot. Have a friend go to another hill and talk to you on the radio.

Drive out to the proposed site and see with the binoculars if the shot is clear. Check for trees or buildings that might block the path. The best time to plan a long-distance shot is during spring when everything is wet and growing. If it's the dead of winter, keep in mind that trees will soon grow leaves again.

Using the location of both endpoints, one can calculate a bearing and tilt angle to point the antenna. Most GPSs have a function to do this built in. High-gain dishes are more difficult to aim the farther out you go. Take both dishes and roughly point them toward each other. Transmit a signal into one dish. With 802.11 gear, link test software enables you to send a series of management frames.

You can take the output of the other dish and feed it into the spectrum analyzer. You will see a display of frequency (across) versus amplitude (up and down). Pick the channel that has the least amount of noise.

Once you get the antennas close, you will see a spike on the frequency that the transmitting dish is tuned to surrounded by noise. Sweep the antenna on each end one at a time and lock down the antenna at the point where the signal is the strongest. At this point, you should have a sufficient SNR ratio to receive the signal with a sufficient margin.

The 5 and 10 dB attenuators can be used in line to check to see if the link margin is adequate. With 15 dB of attenuation in line, a link should last easily for a few hours. If not, you need to plan on larger dishes and amplifiers.

Indoor Site Survey To perform an indoor site survey, prepare by getting the floor plans of the structure. During an initial meeting, find out which areas need to be covered and which ones don't. For example, do you need coverage in locker rooms and bathrooms? Also obtain information about where existing wiring closets are located and if the wiring closets or hub rooms will be used to connect the APs to the wired network.

Then walk through the structure while looking for existing RF sources. If RF sources exist, note which channels the interference is on and the relative signal's strength. Typical cells can cover closed-off areas such as four classrooms or a large area like a basketball gym or a bowling alley.

Starting at the most complicated area, place a potential AP that has to be within 300 feet of the nearest wiring closet, with a cabling path between the two. Then using a laptop with a site survey tool, find the points where 20 dB SNR is observed. This becomes the cell boundary. Place the trial AP so that its 20 dB SNR cell boundary overlaps the one identified by the first trial location. Continue to lay out cells until the whole structure is covered.

How to Make a Frequency Plan

After completing an RF site survey, you'll have a good idea of the number and location of APs necessary to provide adequate coverage and performance for users.

2.4 GHz Frequency Reuse

The 2.4 GHz band has 11 22-MHz-wide channels defined, starting at 2.412 GHz every 5 MHz through 2.462 GHz. Three nonoverlapping channels are available: 1, 6, and 11, as shown in Figure 7-6. These nonoverlapping channels can be used in a three-to-one reuse pattern, as shown in Figure 7-7.

5 GHz Frequency Reuse

The operating channel center frequencies are defined at every integral multiple of 5 MHz above 5 GHz. The valid operating channel numbers are 36, 40, 44, 48, 52, 56, 60, 64, 149, 153, 157, and 161. The lower and middle U-NII subbands accommodate eight channels in a total bandwidth of 200 MHz. The upper U-NII band accommodates

Figure 7-6
2.4 GHz has three nonoverlapping channels.

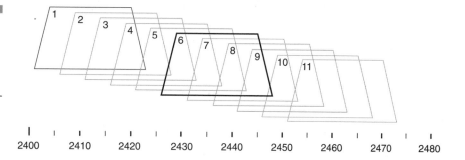

Figure 7-7
Three-to-one reuse pattern

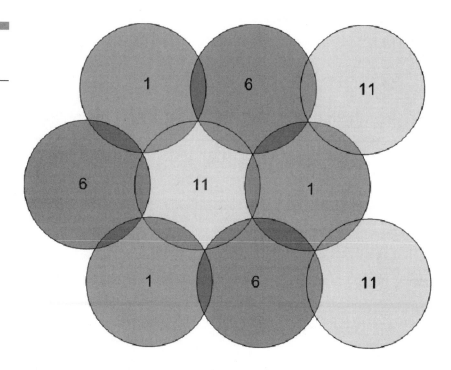

four channels in a 100 MHz bandwidth. The centers of the outermost channels are 30 MHz from the bands' edges for the lower and middle U-NII bands, and 20 MHz for the upper U-NII band (see Figure 7-8).

Point-to-point links operate on the other four channels: 149, 153, 157, and 161. This allows four channels to be used in the same area.

802.11a APs and client adapter cards operate on eight channels: 36, 40, 44, 48, 52, 56, 60, and 64. This allows two four-to-one reuse patterns to be used (see Figure 7-9).

Figure 7-8
5 GHz channels

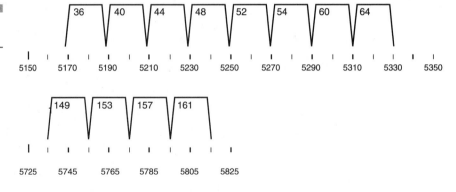

Figure 7-9
Four-to-one
reuse pattern

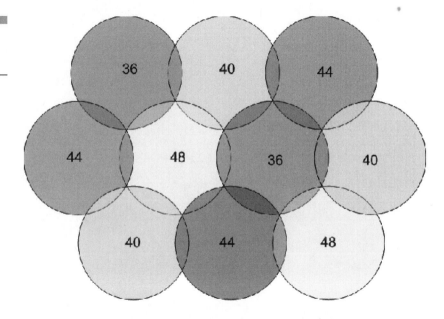

Using both the low- and mid-frequency ranges together allows a seven-to-one reuse pattern with a spare. The spare can be added for a fill to extend coverage or to add capacity in areas like conference rooms where more capacity is needed (see Figure 7-10).

Frequency Allocation

For a simple project like one or two APs, simply assign the least used frequencies from the site survey. For more complex projects involving

Figure 7-10
Seven-to-one
reuse pattern
with a spare

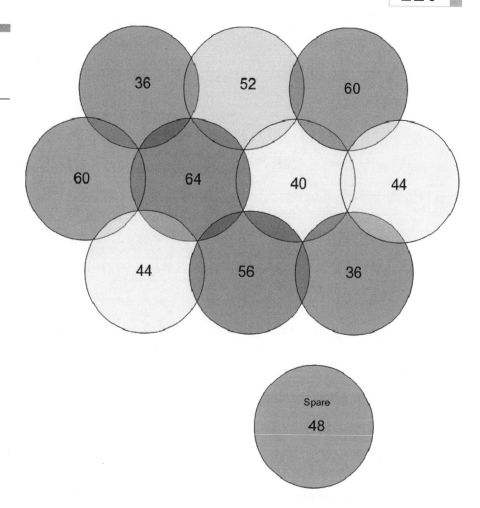

three or more APs, pick a frequency reuse pattern for the frequencies that are used for the project; start with the most complicated part of your site survey and start assigning frequencies. Plan the location of APs initially for coverage, not capacity. Avoid overlapping channels if possible. However, if an area has to be overlapped, plan it such that it is naturally an area where the most capacity would be required, such as in a library, conference room, or lecture hall (see Figure 7-11).

For multiple-floor installations, if more than 30 dB of isolation is used between floors (such as concrete and rebar floors), try not to use the same frequency directly above or below a cell that has already

Figure 7-11
Overlap in a
central library

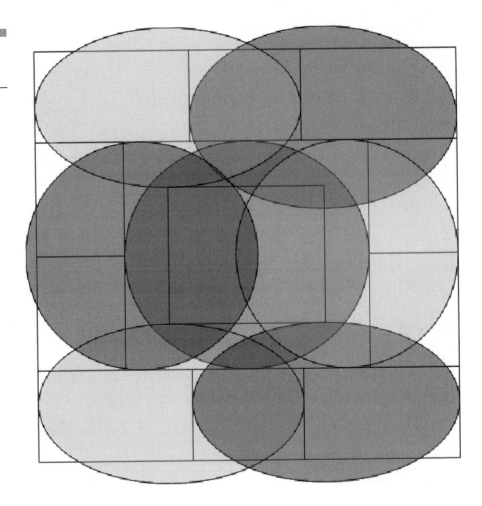

been allocated. Where less than 30 dB of isolation is used between floors (such as a two-by-four framed apartment building), the plan needs to take the three-dimensionality of the cell into account (see Figure 7-12).

Some of the most complex problems are areas where not enough channels are available to plan out the space. In a two-dimensional space, this can happen in areas where a central room has to be covered with surrounding classrooms or offices, such as a library or lab. In a three-dimensional space, this can happen in a tri-level portion of a building (see Figure 7-13).

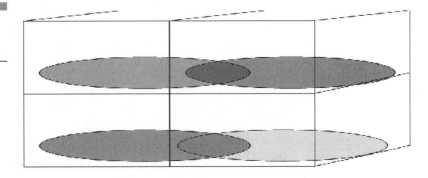

Figure 7-12
Cells are 3-D in
buildings.

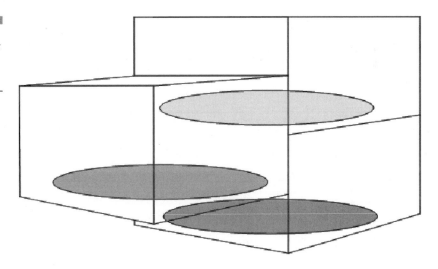

Figure 7-13
A case where RF
planning is
difficult

The signals from different floors can overlap and intrude on another. In some cases when you have no way around two cells that use the same frequency being bordered against each other, plan the seam to be in areas where no coverage is necessary, such as equipment rooms, restrooms, wiring closets, stairwells, and janitorial supply rooms.

Equipment Selection

Now comes an important phase in network deployment; selecting the equipment that is right for this network. Inevitably, the process is an equation balancing costs vs. capabilities vs. requirements.

How to Pick the Right 802.11 Client Adapter Card Depending on what kind of computers you have, several choices of Wi-Fi adapters are available. Figure 7-14 shows the choices of client adapters available.

For a desktop computer, basically two choices are available: an internal or an external installation. The easiest installation is an external Wi-Fi adapter. The installation is really simple; the unit plugs into the *Universal Serial Bus* (USB) port. The disadvantage of this type of product is that in public spaces the PCMCIA card tends to wander off if not properly secured. Some external Wi-Fi adapters have a little oval hole that can receive the same kind of cable used to secure laptops.

For the internal installation, two different products are available. One is an adapter that takes a standard PCMCIA card and adapts it to a PC card adapter that fits in a PC slot. This type of adapter enables the user to remove the PCMCIA card and use it in another

Figure 7-14
802.11 client
adapter choices

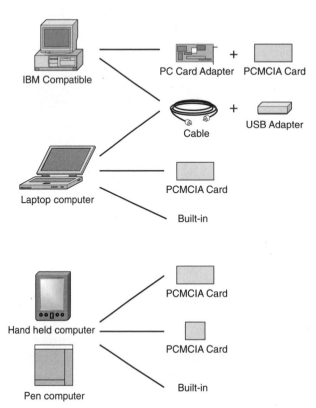

computer. Like the external USB Wi-Fi adapters, the PCMCIA card may get stolen if not properly secured. Another product that is a PC card has all of the radio equipment built in. Typically, the antenna attaches externally to the PC card with a microwave connector. The major disadvantage of the internal installation is that it takes some PC installation skills, and many of the adapters will not work on PCs that run Apple operating systems.

Laptops offer three choices. The most common approach is to use a PCMCIA card. However, more and more laptops have a Wi-Fi adapter built in. One can also use an external USB Wi-Fi adapter, a PCMCIA card, or a built-in Wi-Fi. An external USB adapter is difficult to carry in a laptop bag due to its size. However, it can be used with a variety of computers. Handhelds and tablets can use a PCMCIA card or a *Compact Flash* (CF) card, or have a Wi-Fi adapter already built in.

More and more computers are shipping with built-in Wi-Fi. Considering that a built-in Wi-Fi adapter does not stick out of the laptop where it might break off or have to be inserted every time it needs to be used, it's much more convenient to have it built in. However, one needs to take a careful look as to the upgradeability of a built-in adapter. In the near future, 802.11 will be correcting the flaws with its security protocol, *Wired Equivalent Privacy* (WEP). This upgrade will require more processing power. Also, a whole alphabet soup of new protocols are currently being developed. Therefore, a laptop with a built-in Wi-Fi adapter without upgrade capabilities may quickly become obsolete.

Another thing to consider is that many of the tools for Wi-Fi are chipset dependent. Examples of these are roaming clients, network sniffers, and protocol analyzers. These typically will only run certain brands of cards. For example, Boingo's roaming client will work only with a certain subset of *network interface cards* (NIC). Check the compatibility of prospective NICs using the chart at www.boingo.com.

How to Look at the Specs   Perhaps the most important spec to consider when looking for 802.11 equipment is receive sensitivity. This is the signal strength required for the card to overcome channel noise. A good receive sensitivity (a lower dB number) means little signal sensitivity is needed to acquire the signal. For example, a receive sensitivity of -86 dBm may be okay, but a receive sensitivity of -91 dBm is better. Usually, this figure is part of the specifications. If it is not listed, then usually it's not worth bragging about. Freenetworks .org offers a reasonably current and inclusive chart of the receive sensitivity for various cards.

The next figure to look for is transmitter output. This is expressed in mW or in dBm. Typically, a transmitter will have an output between 20 and 100 mW (or 13 and 20 dBm). It is desirable to be able to control the output power so that interference issues can be mitigated. The combination of receiver sensitivity and transmitter is a major contributor to range.

It is useful to have removable antenna connectors on both APs and client adapter cards. Usually, the connectors are nonstandard, employing either a reverse polarity or a reverse thread from common connector styles. An external antenna connector enables the user to work with antennas to shape where the signal goes. Client cards with external antenna connectors such as the Orinoco Gold card can be connected to desktop antennas such as the Orinoco indoor range extender antenna. This type of antenna can double the range over an ordinary client adapter card.

How to Buy the Right AP   Depending on your application, several choices of Wi-Fi APs are available. First, ask yourself if your application can work with a consumer product or if it needs a commercial product.

Consumer Applications   Residential APs feature low prices, a minimum feature set, and minimal support. However, several configurations can be bought: an AP; an AP and router combination; and an AP, router, and print server combination. If you already have a *Digital Subscriber Line* (DSL) or a cable modem and a router/switch, buy an AP. If you have only the DSL or cable modem, then buy a combination router/switch/AP.

Some AP special features enhance QoS or speed. Some examples of consumer manufacturers are Linksys, Intel, Microsoft, D-Link, Netgear, Belkin, HP/Compaq, and Siemens. If you need more than these features, you will need to look at a commercial AP.

Commercial Applications   A commercial AP offers many more sophisitcated features than a standard AP. Such features include a plenum rating (which enables you to put an AP into the ceiling of a commercial building), *Power over Ethernet* (PoE) (which enables you to power the AP through Ethernet cable), or 802.1x/*Extensible Authentication Protocol* (EAP) (which enables you to authenticate users against a Radius server). Although more expensive, commercial APs have a full feature set, with higher reliability and quality, and more security features. Some examples of commercial manufacturers are Cisco, Lucent/Orinoco/Agere, Proxim, 3Com, and Enterasys.

The WAN Connection

The *wide area network* (WAN) is the limiting part of the wireless network. The Internet backbone has plenty of bandwidth and the Wi-Fi APs have plenty of bandwidth, but the WAN connection to the Internet is bandwidth limited. The choices currently available are

- Fractional DS3
- T1
- Frame relay
- Cable
- DSL
- *Integrated Services Digital Network* (ISDN)
- Wireless

Basically, you get what you pay for.

The *Local Exchange Carriers* (LECs) provide the ISDN and DSL. Bandwidths 128 to 768 Kbps are common. Wireless and cable providers can offer up to 1.5 Mbps connections, but often they are not reliable depending on the *service level agreements* (SLAs) you get. Sometimes repair time is a major issue. If the service fails, the provider may wait until the next business day to repair the problem. These are a low-cost choice and intended for the consumer market. The *Terms of Service* (ToS) usually don't allow you to resell the service. If you get a business subscription, the TOS and SLA may be more agreeable towards a faster repair time, allowing you to resell the connection.

Fractional DS3, T1, and frame relay are all point-to-point services also provided by the LECs and don't come with Internet service. To get Internet service, you need to run the backhaul to an ISP or to a network AP and become your own ISP. QoS would be maintained throughout the network. Of course, T1 and frame relay are priced as a business and thus are available at a much higher cost. The bandwidth of a T1 or frame relay is 1.5 Mbps. Fractional DS3 is an aggregate of several T1s. Bandwidth is in multiples of 1.5 Mbps up to 45 Mbps and is priced accordingly.

How to Extend Range

The range of Wi-Fi systems can be increased with antennas and amplifiers.

Amplifiers   It is best practice to use the minimum power required. However, sometimes the range with off-the-shelf equipment is simply not enough to do the job. It is possible to extend the range of an AP by increasing the transmitter power, but this is limited by the legal limit, the receive link, and, ultimately, interference. FCC regulations limit the transmit power output (refer to "FCC Regulations"). If you use an amplifier to increase the transmit power, the transmit range will be increased, but it will also be necessary to either increase the power of the other transmitter or increase the receive sensitivity of the receiver since the link budgets have to be more or less balanced. Increasing power also has another drawback: It causes interference to other users and does not eliminate sources of interference on the receive side. Both amplifiers used on the transmitter and receiver also have another drawback. Amplifiers blindly increase both the signal and the noise. Usually, amplifiers are used in combination with good antenna systems.

Amplifiers are available for use on the 2.4 to 2.4835 ISM band with a power output up to 1 watt. Models with more power output can be used, but they are only usually sold to licensed amateur radio operators or to export and military accounts. Amplifiers are built for a variety of applications on the ISM band. Thus, it is important to buy one that is suitable or optimized for the modulation (FHSS,

DSSS, and/or OFDM) that is being used. Most amplifiers are usually intended to mount right at the antenna and are often fed DC power over coax with an injector. The amplifier has to have a signal applied to it that is within a certain power level. Feeding the amplifier with too much signal will cause it to have more power in the sidebands or "splatter." With not enough power, the amplifier will transmit too much noise on the signal.

Certain specifications should be considered when buying amplifiers. These include harmonic products (which should be less than 70 dB), the noise figure of the receiver (the lower the better), a fast-acting auto-sense TX/RX switch (which should be fast enough for 802.11), and an AGC receiver (which prevents overloading the receiver in the presence of strong signals).

Antennas Antennas offer another way to increase the range. Antennas limit energy directed in certain areas and redirect the energy in other areas. All antennas exhibit this to a certain extent. A theoretical antenna point source called isotropic is used as a reference for all other antennas. Thus, the gain of an antenna is measured in terms of dBi, which stands for decibels relative to an isotropic radiator or *decibels over isotropic*. Omni-directional antennas generally have between 2 and 10 dBi, whereas directional antennas can have between 3 and 25 dBi of gain.

FCC regulations limit how much gain a transmitting antenna can have (refer to the section "FCC Regulations"), but antennas have two distinct advantages over amplifiers. One, an antenna offers gains in both the transmit side and the receive side. Thus, the impact on the link budget is balanced. The other is that antennas help the interference problem. The transmitter only transmits the signal where it is needed and the receiver only listens to where the antenna is pointed. This has the effect of limiting interference by not transmitting where other users are while receiving more of the intended signal and less of the interfering station (unless of course the interfering station is located in the same antenna path as the intended station).

A number of antennas are available for both APs and client adapter cards with antenna connectors. On the AP side are omni-directional antennas like dipoles and collinears, as well as directional antennas like patch, yagi, grid, and dish antennas. Also, special phased arrays are available that are combined with APs. On the client adapter side are desk antennas, handheld yagis, and magnet mounts.

Antennas: AP Side It can be said that the wireless internet revolution is a silent and invisible revolution except for one thing: antennas. Antennas are the one visible aspect of 802.11. The type of antenna employed determines the range and application involved with a given deployment. The following paragraphs describe the advantages and disadvantages of the different antenna technologies employed in 802.11.

Omnidirectional Omnidirectional antennas transmit their signal in all horizontal directions almost equally. The radiation pattern has the shape of a large donut around the vertical axis, as in Figure 7-15.

The gain is in the horizontal direction at the expense of coverage above and below the antenna. The rubber duckies (the flexible antenna structures on popular access point units) that come with many APs are dipoles that usually have 2.2 dBi of gain. Plenumrated APs with external dipoles work well mounted above ceiling tiles. Some APs have integrated dipole antennas and are suitable for both walls and ceilings.

For more gain, or an outdoor omnidirectional, consider a collinear antenna. Typically, a collinear omnidirectional antenna looks like a *permanent virtual circuit* (PVC) pipe that is between one and five feet tall and has an N connector at the end. Gain for these antennas is between 3 and 12 dBi. A collinear antenna is shown in Figure 7-16.

Vertical This is a garden variety omnidirectional antenna. Most vendors sell several different types of vertical antennas, differing

Figure 7-15
Coverage from an omni-directional antenna

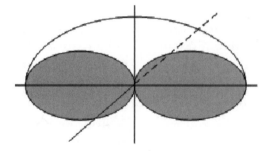

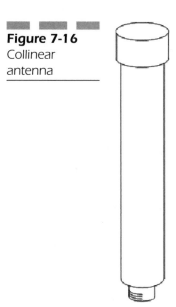

Figure 7-16
Collinear
antenna

primarily in their gain; you might see a vertical antenna with a published gain as high as 10 dBi or as low as 3 dBi. How does an omnidirectional antenna generate gain? Remember that a vertical antenna is omnidirectional only in the horizontal plane. In three dimensions, its radiation pattern looks something like a donut. A higher gain means that the donut is squashed. It also means that the antenna is larger and more expensive, though no antennas for 802.11 service are particularly large.

Vertical antennas are good at radiating out horizontally; they're not good at radiating up or down. In a situation like this, it is better to mount the antenna outside a first- or second-story window. Verticals, like the rubber 2.2 dBi rubber duckies, are in a sense vertical dipoles.

Dipole A dipole antenna has a figure-eight radiation pattern, which means it's ideal for covering a hallway or some other long, thin area. Physically, it won't look much different than a vertical. Some vertical antennas are simply vertically mounted dipoles.

Directional To shape an antenna to a particular part of a building, patch antennas can be used (see Figure 7-17). For example, one may place a patch antenna near the outside of a building to aim the

Figure 7-17
Patch antenna

signal toward the inside of the building. In another case, a patch antenna can be used to limit the signal in an adjoining area. Patch antennas are flat and typically between four and six inches square and have a variety of coax connectors available. They are typically wall mounted. The coverage pattern for a directional antenna looks like Figure 7-18. The gain for a patch antenna is typically between 3 and 15 dBi, and it has a wide beam width.

Sector panel antennas are often used outdoors to cover a sector of a cell. They typically cover 180, 120, or 90 degrees in beam width and have gains between 12 and 20 dBi. These antennas are commonly fitted with an N connector (see Figure 7-19).

Yagi For a point-to-point shot, consider a Yagi antenna. A Yagi antenna is a moderately high-gain unidirectional antenna. It looks somewhat like a classic TV antenna or like washers threaded on a rod. Often, Yagi antennas are mounted inside a PVC pipe to protect them from the weather. A number of parallel metal elements are set

Figure 7-18
Coverage from
a directional
antenna

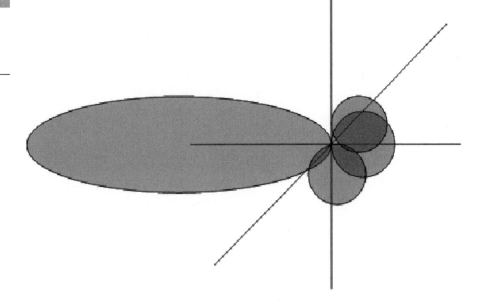

Figure 7-19
Panel antenna

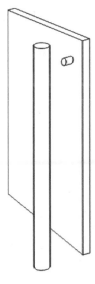

at right angles to a boom. Commercially made Yagis are enclosed in a radome. Yagi antennas for 802.11 service have gains between 12 and 18 dBi; aiming them is not as difficult as aiming a parabolic antenna, though it can be tricky. A Yagi in a radome is shown in Figure 7-20. The gain is fairly high, around 15 to 20 dBi.

Parabolic   For long-distance point-to-point shots, choose a parabolic grid or a dish antenna. This is a very high-gain antenna. Figure 7-21 shows a parabolic grid antenna. Because parabolic antennas have very high gains (up to 24 dBi for commercially made 802.11 antennas), they also have very narrow beam widths. Parabolic antennas are used for a link between buildings. Because of the narrow beam width, they are not useful for providing services to end users. Vendors publish ranges of up to 20 miles for their parabolic antennas. Presumably, both ends of the link use a similar antenna. Front-to-

Figure 7-20

Yagi antenna in a radome

Figure 7-21

Parabolic grid antenna

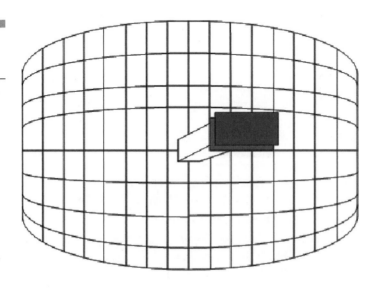

back ratios and wind load are important factors to consider in parabolic grid antennas.

Electronically Steerable Antennas   The best approach to optimize network performance is to maximize the antenna gain to and from the intended subscriber and minimize it elsewhere. Recently, hockey-puck-shaped steerable antennas, six inches across and three inches high, have become available on 2.4 GHz and offer 6 dBi of gain.[9] These can be used with a laptop. Models that can be adapted to APs will follow.

Another company, Vivato, offers an electronically controlled phased array that claims to have a range of up to 7 km to an ordinary client access card.[10] Vivato's Wi-Fi Switches create highly directed, narrow beams of Wi-Fi transmissions. The Wi-Fi beams are created on a packet-by-packet basis. These narrow beams allow simultaneous Wi-Fi transmissions to multiple devices in different directions. Since power is directed only where it is needed, the switch reduces cochannel interference.

Antennas: Client Adapter Card Side   External antennas can be used with laptops to increase their range from APs. These antennas come in two styles: ones that stand on a desk and ones that clip onto the lid of the laptop. They have about a 5 dBi gain and must be bought with connectors that match the external antenna connector of the client access card.

A handheld Yagi antenna can be used to find rogue APs or in a site survey to find sources of RF. They are highly directional, with narrow beam widths and about 10 dBi of gain to make it easy to follow the lead to the RF source. Typically, these have an N-type female connector on them. Adapting them to your client access card requires a short piece of coax that has an N-type male connector and a microwave connector that matches the client access card on the other. This special piece of coax is sometimes referred to as a *pigtail*.

---

[9]See www.paratek.com.

[10]See www.vivato.net.

Another type of antenna is a magnet mount. Unlike the handheld Yagi, it's omnidirectional and has only 5 dBi of gain. It can be used to survey areas for RF sources while driving around (called *war driving*) or it can enable you to utilize hotspots in the convenience of your car.

Antenna Specifications   Table 7-7 shows typical specifications for antennas and how to interpret them.

Gain   The gain of the antenna is the extent to which it enhances the signal in its preferred direction, and it is measured in dBi. An isotropic radiator theoretically radiates equally in all directions. Simple external antennas typically have gains of 3 to 7 dBi. Directional antennas can have gains as high as 24 dBi.

Half-power Beam Width   This is the width of the antenna's radiation pattern, measured in terms of the points where the antenna's radiation drops to half its peak value. Understanding the half-power beam width is important to understanding your antenna's effective

Table 7-7

Antenna Specifications

Frequency range	Should cover at least 2.4–2.4835 MHz.
Gain	Should be expressed in dBi. This figure depends on the antenna.
VSWR	1.5:1 or 2:1 maximum is typical; lower is better.
Polarization	Vertical, horizontal, or circular
Half-power beam width	Degrees for vertical and horizontal. This depends on the purpose of the antenna.
Front-to-back ratio	This depends on the purpose of the antenna.
Power handling	Should handle the transmitter's output power at three times the amount.
Impedance	Should match the transmitter. Usually 50 Ohms.
Connector	N-female is common because it is the strongest, but others are available on commonly available antennas.

coverage area. For a very high-gain antenna, the half-power beam width may be only a couple of degrees. Once outside the half-power beam width, the signal typically drops off quickly, depending on the antenna's design. An antenna's receiving properties are identical to its transmitting properties. An antenna enhances a received signal to the same extent that it enhances the transmitted signal.

Non-standard Connectors   Unlicensed transmitters operating under Section 15.203 are required to be designed so that no antenna other than the one furnished by the party responsible for certifying compliance is used with the device. This can be accomplished by using a permanently attached antenna or a unique coupling at the antenna and at any cable connector between the transmitter and the antenna. FCC Part 15.203 states that intentional radiators operating under this rule shall be designed so that no antenna other than that furnished with the radiator by the responsible party shall be used with the device. The reason for adopting this rule is to prevent the use of unapproved, after-market, high-gain antennas or third-party amplifiers with a device or system.

To meet this requirement, the FCC allows several options. The first option is a permanently attached antenna. These antennas usually include those devices that the box must be opened for in order to remove the antenna. A nonstandard, tamperproof screw secures the antenna to the box, the antenna is soldered to the box, or the antenna is molded into the radio.

The second option is that the antenna be professionally installed. However, the FCC's definition had been somewhat ambiguous. For the most part, high-gain antennas designed to be mounted on a building exterior or a mast generally fall under the professional-installation clause. It's generally understood that a "professional" is one who is properly trained and whose normal job function includes installing antennas. Several groups, including Cisco, the *Certified Wireless Network Professional* (CWNE) Program, and the *National Association of Radio and Telecommunications Engineers* (NARTE), are offering certification programs for unlicensed wireless system installers that would qualify an installer as a professional.

The third option allows a nonstandard or unique connector to secure the antenna to the transmitter. Standard clearly includes connectors such as TNC, BNC, F, N, SMA, and other readily available connectors. The usual convention is that the male connector has a pin in it, and it also has the threads on the inside. More esoteric connectors, such as MCX and MMCX or connectors that are similar to the standard connectors with reversed threads, nonstandard threads, nonstandard shells, or with the gender reversed, are incorporated into Wi-Fi equipment. Some common examples are RP-TNC, RP-BNC, and RP-SMA. Basically, an RP-TNC chassis connector has a male core and a female outside. That is, the threads are on the outside of the connector, but the connector has a pin in it. The mating part has a female core and the threads on the inside. If you need these nonstandard connectors, rest assured they are difficult to find. The best place to get them is over the Internet.

Lightning Protection, Grounding, and Bonding   It is important to properly ground any external antenna. Many volumes have been written about lightning protection, grounding, and bonding. Refer to these and the manufacturer's suggestion. However, if these are not provided, one of the key things to do is to provide an adequate ground through a ground rod. A ground lead should run from the rooftop antenna clear down to the ground rod with a minimum of bends in the line. The Ethernet connection should have a lightning arrestor on it that is connected to the ground system before going into the building. Also, it is helpful to put a loop in the Ethernet cable near the AP or bridge and near where it goes into the building.

RF Propagation   Relative to deploying Ethernet cable to install a wired network, RF propagation can be a difficult science. The following pages describe the engineering challenges related to installing a wireless network, particularly regarding limitations in range.

Multipath Interference   One of the major problems that plague radio networks is multipath fading. Waves are added by superposition. When multiple waves converge on a point, the total wave is simply the sum of any component waves.

When two waves are almost exactly the opposite of each other, the net result is almost nothing. Unfortunately, this result is more common than you might expect in wireless networks. Most 802.11 equipment uses omnidirectional antennas, so RF energy is radiated in every direction. Waves spread outward from the transmitting antenna in all directions and are reflected by surfaces in the area. Figure 7-22 shows a highly simplified example of two stations in a rectangular area with no obstructions.

This figure shows three paths from the transmitter to the receiver. The wave at the receiver is the sum of all the different components. It is certainly possible that the paths shown in Figure 7-22 will all combine to provide a net wave of 0, in which case the receiver will not understand the transmission because no transmission exists to be received.

Because the interference is a delayed copy of the same transmission on a different path, the phenomenon is called multipath fading or multipath interference. In many cases, multipath interference can be resolved by changing the orientation or position of the receiver.

Intersymbol Interference (ISI) Multipath fading is a special case of intersymbol interference. Waves that take different paths from the transmitter to the receiver will travel different distances and be delayed with respect to each other, as in Figure 7-23. Once again, the

Figure 7-22
Multiple wave paths in an unobstructed field

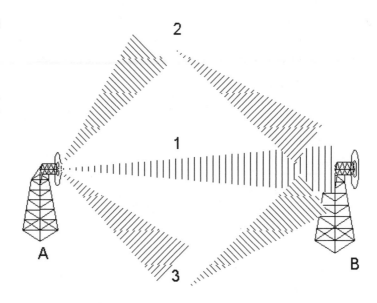

Figure 7-23
Intersymbol
interference

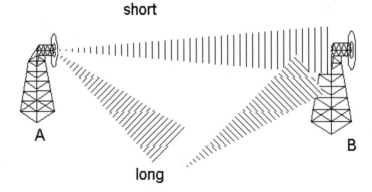

two waves combine by superposition, but the effect is that the total waveform is garbled. In real-world situations, wavefronts from multiple paths may be added. The time between the arrival of the first wavefront and the last multipath echo is called the delay spread. Longer delay spreads require more conservative coding mechanisms. 802.11b networks can handle delay spreads of up to 500 ns, but performance is much better when the delay spread is lower. When the delay spread is large, many cards will reduce the transmission rate; several vendors claim that a 65 ns delay spread is required for full-speed 11 Mbps performance at a reasonable frame error rate. A few WLAN analysis tools can directly measure delay spread.[11]

Using Two Antennas for Diversity   Diversity is often used with cellular base stations to help overcome multipath problems. Many APs have two antenna connectors for diversity, but exactly how is diversity can be implemented besides connecting two rubber duckies has been left to the radio system designer.

Anyone who listens to a car radio while driving in a downtown urban environment has experienced a momentary dropout or fading of the signal at any given stoplight. If you at this point move forwards or backwards ever so slightly, the station comes back in. Although the car is in range of the radio tower, no signal is received

---

[11]Ibid., pp. 158–163.

in these dead spots. This phenomenon is called *multipath fading* and is the result of multiple signals from different paths canceling at the receiver antenna. Figure 7-24 shows multipath cancellation from a large building.

Five different types of diversity can be used to increase signal reception in the presence of multipath fading: temporal, frequency, spatial, polarization, and angular. The first two types of diversity require changes in hardware.

Temporal diversity involves lining up and comparing multiple signals and choosing the one that best matches the expected time of arrival for a signal. This concept is implemented in some digital technologies. One of the most common temporal diversity methods is to use adaptive equalization and RAKE[12] receivers. These have limited use with Wi-Fi of the large DSP MIPS requirements required to handle large delay spreads.

A RAKE receiver is the digital section of a spread spectrum receiver, which can separate out the relevant signal from all the other signals using multiple correlators thereby providing diversity.

Figure 7-24
Multipath

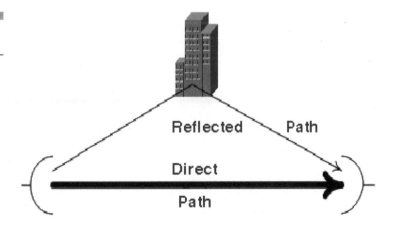

---

[12]The word "RAKE" is not an acronym but is commonly written in all capitals in the literature. RAKE derives its name from its inventors Price and Green in 1958. The multiple correlators from the rake's fingers and the retrieved signal forms the handle of the rake. R. Price and P. Green, "A communication technique for multipath channels," *Proceedings of the IRE*, pp. 555–570, 1958.

Figure 7-25
A Rake receiver

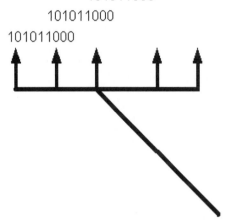

See Figure 7-25. However, the use of RAKE receivers in Wi-Fi is limited if no impractical due to the large DSP MIPS requirements required to handle large delay spreads.

Frequency diversity can be implemented by using two separate radio links on two different channels. If a null exists due to the cancellation of two signals because of a reflection, it will not happen on another frequency at the same place. Routers at both ends of the link can be used to send data across both wireless links. If one fails due to fading, the effective throughput is decreased. The redundancy of the link would also provide protection for other cases of failure.

Spatial diversity helps overcome the multipath problem by using two identical receive antennas separated by a fixed number of wavelengths. If a null happens due to a cancellation of the two signals, it will not happen at the other antenna. Since the antenna with the strongest signal is selected, the link is more likely to survive a fade when using spatial diversity.

The multipath problem can be helped with antennas mounted at different angles to cover the same coverage area. If a signal from one antenna and its reflection are cancelled, then a signal from a different antenna arriving at a slightly different angle will probably not cancel because the phase has changed. Again, since the antenna with the strongest signal is selected, the link is more likely to survive a fade when using angular diversity.

Finally, the multipath problem can also be mitigated by transmitting and receiving using two feed horns that utilize both vertical and horizontal polarization (or clockwise and counterclockwise polarization). When electromagnetic waves are reflected off of flat surfaces, their polarization can change. When the reflected wave and the direct wave combine to form a null, if the wave had been sent using the opposite polarization, no such cancellation would occur. Since the antenna with the strongest signal is selected, the link is most likely to survive a fade when using polarized diversity.

Weatherproofing   It is important to seal all outdoor connections, but it has to be done in such a way that the sealing can be removed if necessary. Use a combination of vinyl-backed mastic tape, heat-shrink tape, and high-quality electrical tape to seal the connector from moisture. Don't use silicon-based products or other spray-on or brush-on weatherproofing materials. They are very difficult to remove.

Locating Your AP to Where No Power Exists

In places where an AP cannot be powered, such as inside a plenum, an attic, or on top of a roof, it costs about $800 to get an electrician to run power per code, in addition to the $200 to run Cat 5 cable from the wiring closet to the AP. A number of commercial AP manufacturers (Symbol, Lucent, and Cisco) have added PoE into their product designs to bring power over the spare pairs of the Ethernet cable to the AP (see Figure 7-26). An injector is located in the wiring closet close to the power outlet. Their APs take power from spare pairs as part of their design. Also, a number of manufacturers are now offering PoE add-ons for most APs in the form of injectors and taps.

PoE also enables you to place the AP much closer to the antenna, thus reducing signal loss over antenna cabling. An Ethernet signal is carried well over Cat 5 cable, but RF signals at 2.4 and 5.8 GHz are heavily attenuated over coax. Also, Ethernet cabling is much cheaper than coax (see Figure 7-27).

Figure 7-26
PoE in an office

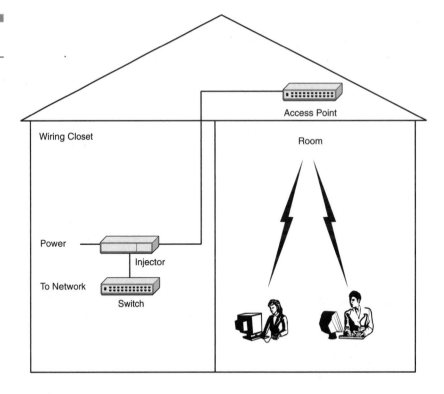

Figure 7-27
PoE in WLAN

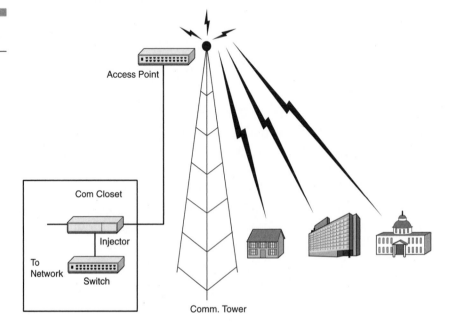

The wiring for PoE is relatively simple (see Figure 7-28). Power is carried over 4&5 and 7&8. However, the polarity differs from one manufacturer to another. Most manufacturers use 4&5 to carry the positive lead and 7&8 to carry the negative lead of the power supply. Cisco's polarity is opposite to what is shown in the schematic.

Besides the polarity, the voltages differ between manufacturer and by model (see Table 7-8). The best practice is to stick with one standard and thus only the vendors of equipment that run on the same voltage and polarity. The IEEE is working on a new standard for PoE called IEEE 802.3af. The new standard will allow equipment from different manufacturers to sense the voltage and polarity of the power that is being supplied on the spare wires of the Ethernet cable and adapt to it.

How to Overcome Line-of-Sight Limitations

The biggest challenge to providing Internet access over a WLAN is line of sight. One of the keys to success as a WISP is to get sites with

Figure 7-28
Simplified schematic of PoE injector and tap

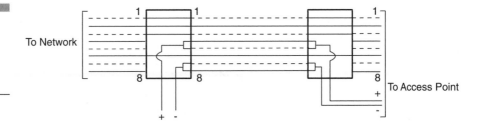

Table 7-8

Injected voltages and polarities by manufacturer

	Pins 4&5 + and Pins 7&8 -	Pins 4&5 - and Pins 7&8 +
5V		
12V	Intel, 3 Com, Symbol, Orinoco	
24V	Intel, 3 Com, Symbol, Orinoco	
48V		Cisco

lots of height above the average terrain or in hilltop locations and link those together using long-haul connections. From the key locations, it's possible to bring the signal to neighboring sites within a thousand feet or so. Also, it's then possible to extend service from one location to a few more as long as redundant paths are brought in to cover the new location. Perfect line of sight is not necessary when the signal is strong enough. It's usually best to set up the wireless backhaul on 802.11a where the noise picture looks better and where the wireless customer won't interfere with the backhaul when installing either 802.11b or 802.11a for the premise (see Figure 7-29).

How to Overcome Interference Limitations

The second biggest challenge to WISP is interference. One of the keys here is to use a band for backhaul that is less congested than the 2.4 GHz band. The 5.8 GHz band fares well for this, except that the range is not nearly as good as the theoretical range on the 2.4 GHz band. Although the 900 MHz band has been touted as being crowded, it may not be as crowded as the 2.4 GHz band. The 900

Figure 7-29
Tiered network
to overcome
line-of-sight
limitation

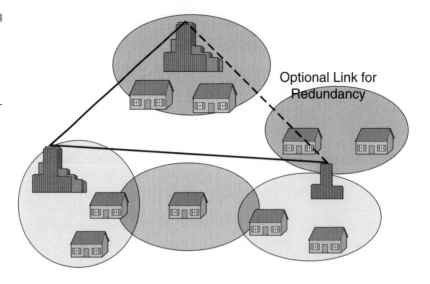

Optional Link for
Redundancy

MHz band has one more advantage: It has much better penetration through vegetation.

Implementation

Now comes implementation time. The following pages will provide guidelines for implementing wireless networks in a variety of settings.

Putting Together a Home Wireless Network

It is fairly simple to set up a home network. The advantage of going wireless at home is that one can use the DSL or cable connection in other rooms besides where the service comes in the house. For example, with a good wireless connection, you can sit outside on a nice day and answer emails and surf the Web. Also, a wireless connection can extend the broadband service to your living room and other parts of the house.

For the most part, it's okay to mix and match equipment from different vendors. The setup is relatively simple. You essentially plug the AP/router into the DSL or cable modem and run a short configuration program. Change your *service set identifier* (SSID) from the default value to something not too obvious. Then set up the client access card in much the same way. Use the same SSID for both your AP and client cards. Once you have it working without the WEP key, the most important part is to use the same WEP key for all the computers in your household and change it often. You can take some other security precautions such as limiting access using MAC filtering.

However, the WEP key department does have some gotchas. Older APs support only 40-bit WEP keys, whereas the new products support both 40 and 128 bit, but it does not stop there. The format of WEP key can be different from one manufacturer to another. Some vendors have you enter a pass phrase and others have you enter ASCII characters. The common denominator is the hex key, so the trick is to use the pass phrase to generate a hex key and use that everywhere else.

Putting Together a Larger Enterprise Wireless Network

In a larger enterprise, two things become a challenge. First of all, more than one AP will be required. This means a site survey and a frequency plan become necessities. Furthermore, APs will need to be of the commercial variety because they will need to be powered over the Ethernet and located in the ceiling. The other challenge is that security becomes more than the small matter of securing Aunt Jane's good jokes and your Quicken books. The company's trade secrets are at stake. A single WEP key now is no longer secure. Other security measures are now necessary, such as 802.1x/EAP and *virtual private networks* (VPNs). It's also necessary to document all your work so that it can be maintained.

Putting Together a School Wireless Network

A wireless school network can help students by bringing the computer lab to the students, rather than bringing the students to the computer lab. This means that computer applications can be used in classes like English, science, and math, rather than being taught as a side course. By having wireless computers, the deployment and setup becomes much simpler. Rather than having to wire every classroom for Ethernet, the school can be set up for wireless. Then about 16 laptops or so with integrated 802.11b cards can be put on a laptop cart that contains a bank of chargers. The advantage of using wireless connections is that students don't trip over the cords. The laptops will last a good morning on a charge. With an extra set of batteries, the laptops can be used again in the afternoon.

The real savings are in the space. The mobile computer lab requires very little room to store. Thus, the computer lab can be used as an extra classroom. The school requires about one AP per four classrooms on average. A typical elementary school requires about 10 APs, a middle school needs about 15 to 20, and a high school between 20 and 30.

The one difficulty with a wireless school network is the two classes of security. It's very important to separate the teacher's network where grades and attendance are kept from the network the students use to access the Internet and store files. However, this does not mean the two networks have to be completely separated from one another. By using Cisco's products, a *virtual LAN* (VLAN) can be set up for the students on a publicly broadcast SSID. It's probably best to make the SSID the same throughout the school district so that students can use their own *personal digital assistants* (PDAs) not only in their schools, but when visiting others in the same school district. An authentication scheme using 802.1x and an EAP scheme can be used to limit the access on the network. Username/password EAPs work well for older students that want to use their own computers and PDAs at school. The lab equipment will work well with 802.1x and EAP/TLS (*Transport Level Security*), which uses a certificate. This way, students don't have to enter usernames and passwords to use the lab machines, but the network is not left open to hackers, spammers, and others wanting to use the wireless connection to launch viruses.

Cisco's products also enable other hidden SSIDs to be associated with other VLANs. Those other VLANs can be routed such that they require authentication using an 802.1x and some EAP scheme. Soon EAP/SIM will be available so that only holders of an *Subscriber Identity Module* (SIM) card or a USB SIM plug will have access to the network. As an alternative, a VPN can be set up for teachers. The added benefit is that they can access their school computing resources from home.

How to Set up a Hotspot out of the Box

Net Near You,[13] Pronto,[14] and Toshiba[15] make it very simple to set up a hotspot for your local coffee shop, hotel, truck stop, or bed and breakfast. The best locations are those that want to promote more customers, that want customers to stay longer, and that are looking for ways to get repeat customers. It's as simple as ordering a DSL or cable connection that enables a resale of the circuit, or bringing in a *wireless*

---

[13]www.nnu.com.

[14]www.prontonetworks.com.

[15]www.csd.toshiba.com.

Figure 7-30
Using a hotspot
in a box

WAN (WWAN) connection from your wireless backbone and installing the hotspot in a box according to the directions (see Figure 7-30).

The hotspot in a box may also contain the AP, or it may simply be a box between the AP and the DSL or cable modem. Depending on what is offered by the network, the box contains a capture portal, authentication software, a firewall, and a router.

The capture portal intercepts the initial web page and puts up an initial login page instead. The authentication software will assist in authenticating the user's username and password. Authentication may also include some identity from a scratch-off ticket or a number given to the user by the barista at the counter by the point-of-sale equipment (similar to the way car washes give you a five-digit number).

If you are a do-it-yourselfer interested in building a community hotspot, check out http://nocat.net. A group in Sonoma County, California, has built an authentication system called NoCatAuth. It makes shared Internet services possible by authenticating participants of the free network to a central authentication server. The server makes it possible to have three different classes of users authenticated by the box. The priority class is an optional class that allows the gateway owner (and anyone else he or she sees fit) priority access to the gateway's resources. The coop class is a mandatory class of service, which gives access to all participants of the cooperative. The public class is open to any unauthenticated roaming client and has the lowest access priority to gateway resources. This class is mandatory and may be defined that the public class is not open to everyone.

With NoCatAuth, a typical logon process goes like this: A roaming user associates with the AP and is immediately issued an *Internet Protocol* (IP) address from the *Dynamic Host Configuration Protocol* (DHCP) server in the NoCatAuth box. The user only has access to the authentication service. The user then opens up a web browser to an arbitrary page. Instead of getting the default page, the user's page is redirected to a secure login page. The user can make one of three choices: Click on a link to find out about the service that is being offered for membership, enter the login information, or click a "Skip Login" button.

If the login procedure is successful, the wireless gateway modifies its firewall rules to grant access to the user and brings up the original web page that the user was trying to access. Another small popup window is opened on the user's display that brings up the login page every few minutes. If the user moves out of range or closes the popup window, the connection is closed.

How to Set up a Hotel or MDU Hotspot

Setting up 802.11 in a hotel or *multiple dwelling unit* (MDU) is a little more complicated than setting it up in a coffee shop. It's likely that more than one AP will exist, the Internet connection must be shared for Internet access, and some wiring inside the hotel is needed. One way to do this is to run fiber from a centralized phone closet to all the floors of the hotel where switches that connect to APs are located, as shown in Figure 7-31. An authentication firewall pre-

Figure 7-31
Hotel or MDU
using fiber

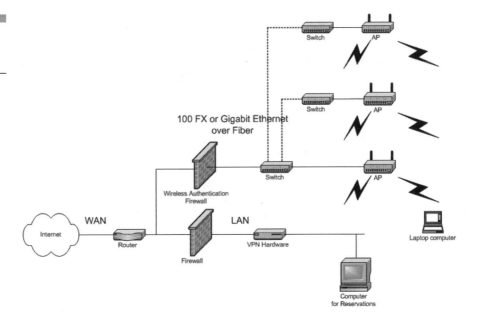

vents access to the Internet until the credentials for access have been presented. The hotel's access to the Internet is separate from the part that provides WLAN service. The hotel has its own firewall to prevent hackers on the Internet or the WLAN from gaining access to the corporate network. The authentication authorization and accounting for this type of deployment are similar to the options in a café-style hotspot. This type of installation works well when it is possible to pull fiber through the building. Installation of the fiber can account for 40 to 50 percent of this project's cost.

However, when it's not possible to pull fiber, consider using the existing or surplus telephone wiring with a *long-range Ethernet* (LRE) router and modems. Figure 7-32 shows the architecture of such a system. LRE works well and is fairly inexpensive. However, many hotels don't have spare telephone lines or have the ability to reuse them.

As a last measure, you can bet the hotel has a *cable television* (CATV) plant. The option shown in Figure 7-33 is the most costly, but it requires very little installation. The broadband router is quite pricey at a cool $20,000 or so. One requirement is that the hotel has a two-way cable system. This means that at worst case all the line

Figure 7-32
Hotel or MDU
using LRE

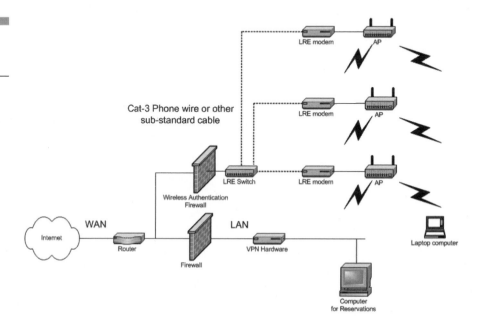

Figure 7-33
Hotel or MDU
using CATV

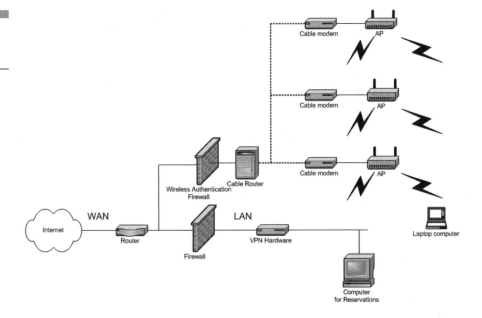

extenders need to be upgraded to two-way. The line extenders can run $1000 apiece.

How to Set up a Layered Office Hotspot

Layering allows an enterprise to share its WLAN for public access. Public hotspots can be created simply by allowing public users to share the WLAN and Internet infrastructure securely. Layering allows the enterprise to sell off its excess bandwidth or to allow public use of their Internet facilities to vendors and contractors without allowing them onto the corporate network. This same concept can be extended into residential areas if the DSL or cable service agreements allow resale of the Internet connection. Two companies, Sputnik and Joltage, have developed layering opportunities within the home network and in *small office, home office* (SOHO) markets.

To enable layering within an enterprise network, one can use an architecture similar to the one shown in Figure 7-34. Only the public SSID is broadcast. However, users can associate with either the public or the private SSID. The public one is used by the public users

Figure 7-34
Using layering
to provide
hotspot service

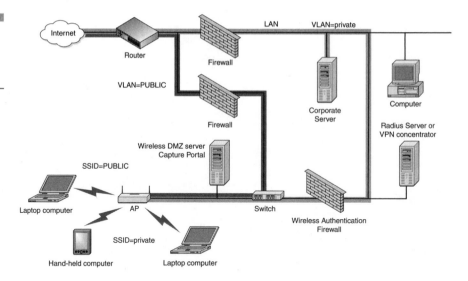

and the private one is used by the enterprise itself. The two SSIDs are linked to their own VLAN. The public SSID links to the public VLAN and the private SSID links to a VLAN that belongs to the enterprise. After being authenticated and authorized, users of the public SSID can access the Internet. In the same way, users of the private SSID can access the corporate network. The switch and router shunt traffic according to its VLAN ID.

Many VLAN systems allow up to 16 separate VLAN IDs. The 802.11 standard currently allows the broadcast of only a single SSID. The rest have to remain hidden but can be associated with the system. If the 802.11 rules could be slightly bent, up to 16 broadcast SSIDs could be supported by sending a different SSID in each broadcast message. This would allow multiple service providers to be supported.

As-Built Documentation

After installing a project, all areas by the installation need to be documented and photographed so that when a modification needs to be done, the As-Built documentation can be used as a reference.

A nice tool called NetViz[16] can be used to document the initial installation and any modification that is done to the network. NetViz is a program that helps you capture your network at all layers. You can import Visio drawings, CAD drawings, text files, and pictures showing rack elevations and installations. The diagrams enable you to drill-down from a nationwide or citywide view all the way down to the IP level.

Operation

Once the network is installed and operating, the network operator has to "Keep 'em flying." This will include dealing with day-to-day challenges, the greatest of which is dealing with dynamic interference issues.

How to Deal with Interference Issues

According to the FCC, all Part 15 users must accept interference as part of doing business over public airwaves, but that does not mean we must accept harmful interference, which is an obstruction of such magnitude that signals carrying other telecommunication services are disrupted. What this means is that normal interference is inevitable; thus, it must be calculated as part of the plan and dealt with as issues come up. Simply cranking up the power causes harmful interference, so other peaceful ways of managing the interference must be found.

A variety of sources cause interference on the 2.4 GHz and 5 GHz bands; some of which are 2.4 GHz cordless phones, Bluetooth-enabled devices, microwave ovens, RF lighting, and other WLANs. Currently, the 5 GHz band is relatively free of interference, but as time goes on, this band too will be crowded.

---

[16]www.netviz.com.

2.4 GHz cordless phones marginally interfere with 802.11b. If it's necessary to buy cordless phones after an 802.11b system is installed, use a 900 MHz model instead.

Bluetooth enables users to link devices wirelessly. Interference from Bluetooth is limited to a short distance due to its limited range. Since Bluetooth devices are used all over the place, they really cannot be anticipated in a design or plan.

Microwave ovens that leak too much can be health hazards. Often the seals fail on older microwave ovens. Sometimes it's worth replacing a heavily used leaking microwave. If it's not heavily used, it only interferes once in a while. Distance from the unit works wonders. It's a good idea to plan APs away from microwave ovens. If an AP has to be close to an AP, plan its frequency on either channel 1 or 11 so that it's far away from the microwave oven's center frequency of 2.45 GHz (which is near channel 6).

RF lighting from Fusion Lighting, a company based in Maryland, uses a standard magnetron similar to the ones used in microwave ovens. The magnetron focuses energy on a quartz sphere the size of a golf ball filled with inert krypton or argon gas and sulfur. The microwave energy excites the gas and effectively lights up a plasma that releases extremely efficient light. The white light created is nearly identical to sunlight and costs much less than fluorescent lighting. Although this application has been approved by the FCC under Part 18, it has the potential of interfering with your Wi-Fi project. Used at North Carolina's Pope Air Force Base and at the Smithsonian Air and Space Museum in Washington, D.C., this type of lighting totally wipes out the use of 802.11b.[17] Most likely, channels 1 and 11 are still usable. Proper planning and a site survey will uncover this.

As you can see, the planning step is very important when it is done methodically on paper because this forces you to think out every step and cover all the bases. Frequency coordination with other WISPs and WLAN users in your area is equally as important

---

[17]Ephraim Schwartz, "From Point B to A," Infoworld, www.infoworld.com/articles/op/xml/02/04/15/020415 opwireless.xml.

as good planning. It's important to start the relationships on a friendly and cooperative basis so that when new links or APs are planned and deployed, they are coordinated together, rather than creating a new crisis. It is important to settle on the same modulation method (FHSS and DSSS) if possible. Communication is probably the most important part of this process.

Maintenance

Once a Wi-Fi system is installed, it still requires periodic maintenance.

Scanning Security Logs

Periodically, security logs should be inspected for unusual activity. Depending on what shows up, actions are required. Security is a continuous process and not a state that is achieved.

Locating Rogue APs

To find rogue APs, one can use a program like Netstumbler[18] or Airopeek[19] with a client access card and a handheld Yagi antenna. Such software and devices can help one spot all the beacons from unknown APs and client access cards in your building. Clues can be found, such as clients using ad hoc mode rather than infrastructure, SSIDs that do not conform to the ones used in the area, and MAC addresses that don't belong to the group of known MAC addresses. Note that the MAC addresses are ones that belong to the air interface and not the MAC addresses of the LAN connection to the network.

---

[18] www.netstumbler.com.
[19] www.wildpackets.com.

Coverage Issues

It is worthwhile to check coverage once in a while. An AP's antenna can become disconnected when air-conditioning or electrical service people climb around in the plenum. It is also possible for the transceivers to cease functioning once in a while. Use the opportunity of looking for rogue APs to check coverage at the same time.

Upgrades

Upgrades are often necessary as technology develops and security enhancements are found. If a new version of AP software is made available, test it in one AP for a week before deploying it to the rest. Once the week passes without incident, deploy the upgrade during a low-service period to avoid service interruption.

Economic Aspects of Wi-Fi

Can using 802.11 applications save subscribers money? Can it make money for service providers? Does it significantly lower barriers to entry to the broadband Internet market? The absence of cabling and obtaining rights-of-way are the first indications of potential savings in the installation of a network. Perhaps one of the strongest arguments in favor of 802.11 is that it presents a potential cost-effective means of offering broadband Internet service to a mass market with the least expense in infrastructure relative to wired technologies (twisted-pair copper, coax cable, and fiber to the home). This low cost in infrastructure promotes the deployment of 802.11 services by less well capitalized entrepreneurs, municipal networks, and even free-net community networks built and maintained by volunteers. The growth of 802.11 networks is often described as being "viral" (that is, unplanned) or "grassroots."

The Economics of Wireless in the Enterprise

The economics of 802.11 in enterprise applications should be assessed in two ways: (1) by comparing applications where the wireless network is simply less expensive to deploy than the wired network where both applications perform the same function and (2) by examining situations where a wireless network enables employees to be more efficient. Money saved is money earned.

You *Can* Take It with You when You Go

802.11 has gained wide acceptance as a technology in enterprise networks. This is because of many reasons, including cost savings, mobility, employee productivity, and possibly layering. The origin of wireless networks rests in the convenience of not having to run *Category 5* (CAT 5) wiring in an enterprise environment. The cost of the wire is not so high; the labor to perform the installation, bore holes in the walls, and deface other property to run the wire increases the cost of a wired *local area network* (LAN) as compared to a *wireless LAN* (WLAN).

Recall the timeless wisdom regarding death and personal wealth: "You can't take it with you when you go." Although Table 8-1 may not point to an overwhelming financial advantage of wireless over wired networks, one point to remember is that enterprise tenants that deploy wired networks *can* take it with them when they go. That is,

Table 8-1

A comparison of installation costs: Wired LANs versus WLANs

Cost Component	Cost per Unit	Number of Units Required for a Wired Network	Total Cost for a LAN	Number of Units Required for a Wireless Network	Total Cost for a WLAN
Cisco 1721 Router	$2,000	1	$2,000	1	$2,000
Cisco 3524 Switch	$2,000	1	$2,000	1	$2,000
Dell server	$2,500	1	$2,500	1	$2,500
Laptop with built-in 802.11	$1,500	10	$15,000	10	$15,000
Desktop 802.11 card for a PC	$1,000	1	$1,000	1	$1,000
Printer	$2,000	1	$2,000	1	$2,000
Cisco 350 Access Points	750	0	$0	2	$1,500
Virtual private network (VPN)/ encryption	$1,500	0	$0	1	$1,500
T1	$500	1	$500	1	$500
Installation CAT 5 wire drops/runs	$250	10	$4,000	3	$750
Totals			$29,000		$28,950

Note: WLAN equipment pricing may fall faster than LAN gear as technology matures.

most commercial lease agreements in North America hold a proviso that wired infrastructure must remain in the building when an enterprise tenant vacates the premises (most do so for more advantageous rent). This is a sunk cost that enterprise tenants lose when they move to another building space. A wireless network, on the other hand, is something that is largely, but not entirely, portable. The deployment of a wireless enterprise network gives enterprises greater flexibility when they are shopping for more advantageous rents because they can take their LAN with them when they go.

Moves, Adds, Changes

Wired enterprise networks are expensive when it comes to moves, adds, or changes for employee seats. For a LAN, the estimated cost for a MAC is upwards of $150 for a computer on the network. When circuit-switched telephony is included, the cost can climb to $500 per seat. As a WLAN recognizes almost any device at almost any location on a network, the costs associated with moves, adds, or changes are largely eliminated.

The Productivity Benefits of WLANs

A recent study conducted by Sage Research revealed that the economic reward of 802.11 networks may not come from a comparison of the costs of infrastructure. Instead, the economic reward comes from the improved productivity of employees using wireless technologies over those who use wired networks. Through interviews with actual WLAN users, this study uncovered WLAN productivity benefits—particularly time savings, flexibility, and quality of work. The following sections detail each of these productivity benefits.

Time Savings In today's high-pressure world, no one can dispute that time is money. The strongest, most quantifiable benefit to using WLAN is time savings. Based on primary research, actual WLAN users indicate that, on average, a WLAN user can save up to eight hours per week versus a wired LAN user. Of course, time savings equates to monetary savings. Using these figures, on average, a WLAN user saves his or her organization $260.50 per week due to time savings.[1]

WLANs enable users to save the most time when responding to e-mail. This is especially true for IT professionals. In fact, an employee within the IT department could receive over 100 e-mails each day. By accessing the WLAN while in meetings, in the cafeteria, or at other locations, employees are often able to catch up on their e-mail while doing other things.

Not only does the WLAN user save time, but other hard-wired employees also save time indirectly. For example, a mobile employee may not need to call a colleague to obtain information. Instead, the mobile user can access the LAN via a wireless connection and obtain the information. Likewise, a nurse working on one floor of a hospital may need access to patient information. Instead of calling the administrative assistant or receptionist, the nurse can access the patient's file on the LAN by using a wireless device. As discussed in the following two sections, by using a WLAN, both the quality of work and flexibility improve in this scenario. These soft benefits that drive WLAN adoption often cannot be quantified.[2] Table 8-2 provides some examples of organizations that are saving time by using WLANs.

Flexibility   Another benefit to deploying WLANs is the flexibility they offer users. WLAN users are able to access the LAN from a variety of traditional areas inside their building(s) as well as nontraditional areas—sometimes even outside their building(s). Table 8-3 shows the various locations, in order of the most common to the least common, where users access their organization's LANs wirelessly.

---

[1]Interviewees were asked to estimate the amount of time WLANs save an average user per week. Interviewees were also asked to estimate the average annual salary of a WLAN user. Based on this data, the average WLAN user saves $260.50 per week in time savings, although savings ranged from $30 per week per user to $750 per week per user. This data is for directional purposes only and should not be considered statistically significant due to the small number of interviews that were conducted.

[2]"Wireless LANs: Improving Productivity and Quality of Life," a white paper from Sage Research, www.sageresearch.com, May 2001.

Table 8-2

Time and
money saved
using wireless
networks

Organization Type	Types of WLAN Users	Average Time Saved	Average Money Saved
Government	Executives and IT	5 hours a week per user	$150 a week per user
Healthcare	Nurses	15 hours a week per user	$472 a week per user
Finance	Executives and IT	15 hours a week per user	$750 a week per user

Source: Sage Research

Table 8-3

Locations
where 802.11
access
enhances
employee,
visitor, and
vendor
productivity

WLAN Access Inside the Premises	WLAN Access Outside the Premises
Conference room	Home
Offices	Another job site
Meeting room	Warehouse
Shipping and receiving room	Train
Inventory area	Bus
Distribution center	Library
Cafeteria	Airport
Training room	Athletic field or gym
Classroom	Restaurant or coffee shop
Vendor lobby (guest access)	Hotel

Source: Sage Research

This broad coverage gives users the flexibility to work at their
desks or move unencumbered throughout the organization while
remaining fully linked to the network. It should be noted that WLAN
deployment varies from organization to organization.

Quality of Work   In addition to flexibility and time savings, many users report that the quality of their work has improved by using a WLAN. The greatest benefit on this front is accuracy. For example, let's say a government employee is responsible for keeping track of drums of nuclear waste. In the days before WLANs, a scientist would need to count the drums by hand and record the number on paper. An administrative assistant would then enter the data into a server-based application at a later date. This produces an incredibly large margin for error. Now, with access to the WLAN, the scientist can use a handheld scanner to read bar codes that are placed directly on the waste drums. By doing this, a great deal of data is fed in real time directly to the server. This has increased accuracy tremendously—instead of knowing where 70 percent of the waste is located, they are now able to know where 99.9 percent of the waste is located.[3]

Another Economic Aspect of Wireless Networks: Vo802.11 in the Enterprise

Efficiencies introduced into a workplace by wireless networks should not be limited to data applications. Vo802.11 introduces some aspects that also save significant time and effort in the workplace, as discussed in the following sections.

Responsiveness to Customers   Customers inevitably expect quick answers to their questions about order status and desire flexibility in responding to changes in orders. Companies with Vo802.11 telephones (one of many 803.11 applications) have found that those telephones improve their service to customers. One method of documenting this value is to estimate the value of retained customers. Companies often know the average profit per customer per year. Through interviews with employees, they estimate the number of customers they anticipate retaining as a result of improved responsiveness to customer concerns. Often, if they can retain one or two customers per year, they will achieve significant dollar savings. This is illustrated in the following example:

---

[3]Ibid.

2 customers retained a year×$5,000 average profit per cus-
tomer = $10,000 savings

Many companies do not have this high average dollar profit per
customer. However, those companies usually have many more customers, so their estimate of the number of customers retained tends
to be higher. See the following example:

1 customer retained a week × 52 weeks a year × $100
average profit per customer = $5,200 savings per year

Supervisor Time Savings It takes supervisors a significant
amount of time to get to an available wall or desk phone within large
facilities. For example, 20 seconds lost walking both to and from a
wall phone is not a lot of time, but when it happens 20 to 30 times
per day (not unusual), it adds up. Companies have conducted time
and motion studies like the following:

20 seconds to and from the wall phone = 40 seconds saved
per call

25 calls per day × 40 seconds saved per call = 1,000
seconds saved per day

1,000 seconds saved per day per 3,600 seconds a hour =
.278 hours per day per supervisor

.278 hours per day per supervisor × 8 supervisors = 2.22
total hours saved per day

2.22 hours saved per day × $15 salary per hour = $33.30
saved per day

$33.30 saved per day × 300 workdays per year = $9,990
saved per year

In this example, $9,990 is saved per year just because supervisors
did not have to find an available wall phone.

Efficiencies in the Maintenance of the Production Line
Maintenance personnel are some of the most enthusiastic users of
Vo802.11 telephones. Vo802.11 telephones enable them to be notified

and respond to production line malfunctions immediately. They also enable the maintenance supervisor to repair machinery using both hands while receiving instructions on the Vo802.11 telephone directly from the machinery manufacturer's technical specialists. This eliminates errors and wasted time spent walking to and from a wall phone after each repair instruction. The continuous operation of production lines is vital to companies. Most companies know what the value of production is per hour. For example, one manufacturer claimed that if the production line goes down, the company loses $1,000 of production per hour. To attach specific dollar savings, supervisors asked their maintenance engineers to highlight specific instances where the Vo802.11 telephones have been utilized and then estimated the number of minutes saved. Their valuation of this benefit looked like this:

> The minutes saved per repair = 15 minutes = .25 hours
>
> The frequency of repairs where wireless telephones made a significant difference = 1 time per week
>
> The value of production per hour = $1,000 per hour
>
> The dollar value of efficiencies = 25 hours $\times$ 1 time per week $\times$ $1,000 per hour = $250 per week $\times$ 52 weeks a year = $13,000 total savings per year

Vo802.11 Telephone System Cost Justification in the Workplace

The following paragraphs describe how Vo802.11 saves enterprises money. Vo802.11 makes managers more efficient and thus saves their employers money.

Cost Savings from Immediately Answering Calls from Long-Distance Customers   When customers or suppliers are statewide or national in scope, their long-distance telephone bills are significant. By immediately answering incoming calls from customers and suppliers, companies save the cost of returning those long-distance calls later. The following calculations illustrate this cost saving:

Average dollar cost per long-distance call = \$2.50

The number of long-distance calls per day per supervisor not made due to the use of SpectraLink Wireless Telephones = 2

The number of supervisors = 6

The number of workdays per year = 300

Dollar savings = \$2.50 $\times$ 2 $\times$ 6 $\times$ 300 = \$9,000 per year

Savings in Interoffice Telephony: A Case Study of Amber-Waves WISP Vo802.11   AmberWaves is a WISP in northwest Iowa. One of their clients has 3 offices with 35 employees, which are linked by an 802.11 network. The greatest distance between the offices is 19 miles. Calls between the offices are long distance, which retails at 10 cents per minute in this rural community. Data circuits are not inexpensive here either.

AmberWaves has equipped customers with 802.11 wireless bridges to connect the LANs in the three offices. By adding a *voice over Internet Protocol* (VoIP) gateway, the company was able to route their interoffice phone traffic onto the 802.11 network between the three offices. This meant all traffic that previously went over the *Public Switched Telephone Network* (PSTN) was diverted to their internal 802.11 network, saving them money on both local phone bills (no need for a large number of lines to serve the 35 employees) as well as eliminating their interoffice long-distance expenses.

This wireless network enables the firm to be its own internal data and voice service provider. The use of Vo802.11 frees the firm from local and long-distance telephone bills (see Figure 8-1). The end users report that the *quality of service* (QoS) on the network is better than the frame relay circuit they previously used.[4]

It is useful to apply Figure 8-1 to Table 8-4 to see how this company has saved money on their telecommunications expenses in bypassing local telephone service providers.

---

[4]Brent Bierstedt (CTO of AmberWave Communications), telephone conversation, November 20, 2002.

Figure 8-1

Using Vo802.11 in interoffice telephony saves on both local and long-distance phone bills.

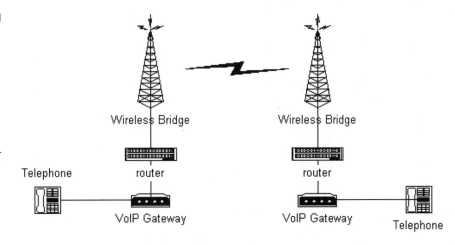

Table 8-4

Worksheet for determining potential savings in bypassing local and long-distance telephone service providers

Number of Phone Lines (T1 and DS3) per Office That Could Be Replaced with Vo802.11	Cost per Month of Phone Lines That Could Be Replaced by Interoffice Vo802.11	Total Savings per Office per Month
Office A		
Office B		
Office C		
Total Savings		

Enterprise Conclusion

One can save money in many ways in an enterprise environment using 802.11. Although many economists might focus on the hard and fast comparison of wired versus wireless, the real savings initially appear less tangible, making a *return on investment* (ROI) or *net present value* (NPV) analysis a little more challenging. When employee mobility and efficiency are taken into account, the savings quickly add up. In situations where Vo802.11 can be employed, an even more tangible financial case for deploying 802.11 can be made when comparing the costs of wired versus unwired phones on the corporate LAN or *wide area network* (WAN).

The Economics of Wi-Fi in Public Networks

What is the economic pull to grow wireless public networks? The previous section of this chapter described the advantages of 802.11 in private networks. How then will public networks grow into acceptance in our economy? In an ideal world, some form of ubiquitous wireless coverage would extend to at least every residence and small business in a metropolitan area. After achieving that goal, extending the coverage to small towns and farms could occur at a rapid pace, assuming a business model propels that growth.

Although the Telecommunications Act of 1996 was intended to bring competition to the local loop, some six years after its passage, less than 10 percent of U.S. residences enjoy any choice in their local telephone service provider. Competition will never come *in* the local loop, but rather *to* the local loop. The act prescribed a formula for competitors to lease facilities (copper wire and switch space) from incumbent service providers. One of the reasons for a lack of competition in the local loop is simply the cost of deploying competing strands of copper wire.

According to studies performed by the *Federal Communications Commission* (FCC), the cost to install a copper loop plant depends on the density of households in the service area. It can range from $500 per household in the cheapest urban sites to a typical $1,000 in dense suburban areas and ascend to $10,000 a loop in outlying rural areas. Economies of scale apply here. A competitor cannot come close to matching incumbent costs on a loop plant, because a competitor with a low market share has, effectively, rural density (and costs) even in an urban area.[5] A competitor to an incumbent telephone company must then consider its ROI on a per-customer basis. If the competitor realizes $40 per month, for example, on a customer, the ROI period could be very long. If a wireless service provider could persuade the customer to purchase his or her own *customer premises equipment* (CPE), the wireless competitor could potentially be more

---

[5]Fred Goldstein (telecommunications consultant), interview, November 28, 2002.

competitive than any other form of competitive service compared to the incumbent telephone company.

The following sections explore how 802.11 could propagate from enterprise networks in dense population centers to public networks and stretch to even the most rural areas. The current debate is whether that propagation will occur top down (that is, as deployed and operated by monolithic, century-old telecommunications carriers) or from below with "mom 'n pop" entrepreneurial operators providing broadband Internet access to their neighbors, schools, and libraries. Community networks pose yet another revolution-from-below scenario where broadband is provided free or nearly free of charge by community volunteers.

Hot Spots

At the time of this writing, the focus of the 802.11 industry is on *hot spots*—that is, locations where laptop-equipped Internet surfers are likely to congregate for the short term (a coffee shop or airport) or reside for the long term (a hotel or *multiple dwelling unit* [MDU]). This is largely a function of the limited range of off-the-shelf *access points* (APs) (50 to 100 meters depending on conditions), access to a broadband Internet source, and a ready power source. In order to be commercially viable, a hot spot must incorporate all of these elements in one time and space. According to Allied Business Intelligence, "WLAN is extending its domain beyond the home and enterprise and is rapidly growing in popularity for public hot spot applications. The need for data connectivity on the go is being spurred by the increasing number of notebooks and PDAs that have WLAN adapters. North American WLAN public hot spot subscriber revenues are expected to increase to $868 million by 2006."[6]

---

[6]Allied Business Intelligence, "Wireless LAN Public Hotspots: Assessment of Business Models, Service Rollouts, and Revenue Forecasts," www.wirelessreport.net/wireless-networks/july02/layeringtocreatepublichotspots.html.

Figure 8-2

Wi-Fi deploy-
ment in a cafe
or coffee shop
hot spot,
Source: Pronto
Networks

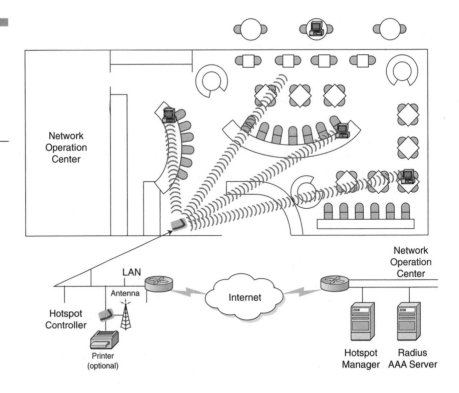

Cafes (Coffee Shops) Wireless hotspot technology allows a cafe-owner to generate additional sources of revenue by offering customers wireless broadband Internet access. Offering wireless Internet access encourages customers to stay longer and spend more, allowing the cafe to capture more revenue per customer. One example of this is T-Mobile's relationship with coffee shop chain Starbuck's. T-Mobile is installing hotspots in thousands of Starbuck's worldwide. Competitors to both Starbuck's and T-Mobile are installing hotspots in coffee shops.

Airports Airports were perhaps the first and most obvious place for hot spot technology. Business travelers needed to access their e-mail and web sites while transiting through airports. Dial-up Internet access was and still is inconvenient because a user must either connect to a pay phone or have access to an airline's private club (however, American Airlines' Admiral's Club now offers T-Mobile's 802.11 service) where dial-up access is available. The deployment of wireless service enables the business traveler to access e-mail and web sites from almost any location in the airport (including aircraft).

Airports have also lost an important source of revenue with the advent of cell phones—pay phone revenue. Business travelers no longer dash to the banks of airport telephones to make calls and check voice mail. Rather, they use their cell phones for which the airport earns no revenue. The deployment of airport hot spots could serve to generate the revenue lost with the declining popularity of pay phones (see Figure 8-3).

Hotels Like airports, hotels have derived revenue, in one form or another, from business travelers seeking access to e-mail and web sites. Business hotels must double as offices on the road for business travelers. This means convenient broadband Internet access must be available in order to remain competitive. Efforts to run CAT 5 wire through business hotels have encountered difficulties because of the

Figure 8-3
Wi-Fi
deployment in
an airport,
Source: Pronto
Networks

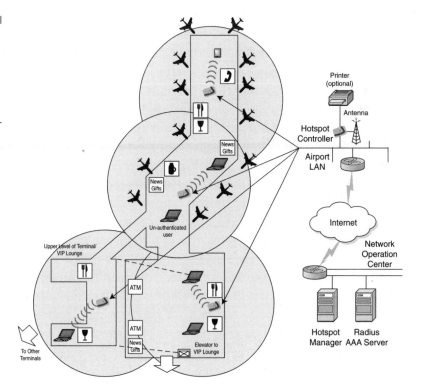

high cost and potential inconvenience of hotel guests not being able to gain access to the network on the first try. Providing Wi-Fi for guests spares the hotel operator the expense of running CAT 5 wire throughout the hotel structure. Users should also find gaining access to the network more convenient on Wi-Fi than they would on wired Ethernet. Like airports, hotels have seen their revenues generated from guest telephone calls drop off from the increased cell phone use. The deployment of a hot spot in a hotel could offset that lost revenue (see Figure 8-4).

MDUs MDUs include apartment buildings and condominiums or other high-density dwellings. The attractiveness of the MDU market is that subscribers are concentrated on one space and can be served with a minimum number of APs (see Figure 8-5).

Turnkey Operators

A promising approach to spreading the coverage of 802.11 is the turnkey of 802.11 services via firms that provide local operators with back-office services to overcome the chief barrier to entry for this market-customer service and network administration. Similar to the early spread of dial-up Internet access services, local operator entrepreneurs can get into this market space for only a few thousand dollars. In the current context of these plans, the local operator offers service in hot spots and shares revenue with the service provider for every per-visit or monthly subscriber. The hot spot operator must, in most but not all cases, purchase the AP hardware and associated access software. The challenge for the local operator is to operate in enough hot spots to be economically viable.[7] Table 8-5 lists turnkey operators, their products and business propositions for local operators. This presents an opportunity for the entrepreneur to enter this market without the expense of maintaining a large engineering staff.

---

[7]"Making Wireless Pay," *The Economist*, www.economist.com/business/displayStory.cfm?story_id=1067140, April 4, 2002.

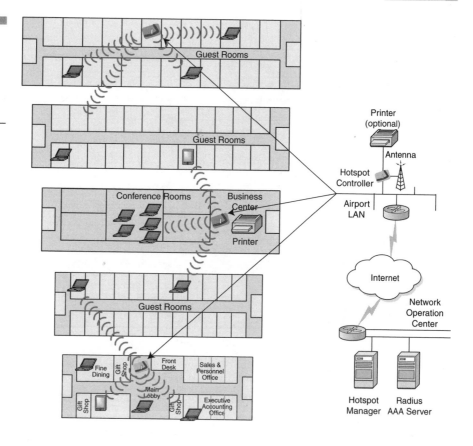

Figure 8-4
Wi-Fi deployment in a business hotel,
Source: Pronto Networks

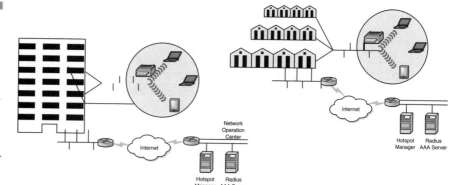

Figure 8-5
Wi-Fi deployment in an MDU, also known as apartments or condominiums,
Source: Pronto Networks

Table 8-5 802.11 turnkey operators, and their product and business models for local operators

Network Operator	Description	How to Make Money
Boingo	This is a hot spot service. Boingo requires the purchase of a $695 Colubris AP configured for authentication, as well as high-speed service and a static IP address. Boingo also offers WISP in a Box, which enables a hot spot to have its own network customers while also offering Boingo service.	Boingo pays a $20 bounty for each new member, and $1 per connection session at a hot spot location. For WISP in a Box, the revenue sharing is the same, but Boingo requires that its partners in this service charge no less than $12 per month, $6 per day, and $3 per hour.
AirPath	AirPath has several plans, including the AirPath Instant Hot Spot. AirPath also allows Boingo Wireless customers to roam on their network.	They offer a variety of equipment, starting at $695, and also have approved equipment that can be deployed. The advantage of their system is that the local operator keeps all the revenue from the local subscriber (from a single point or a network of hot spots) and receives partner revenue from other AirPath members using the point.
NetNearU	This firm offers several kinds of preconfigured hot spot services, but has little detail on their online site. Boingo Wireless subscribers can use NetNearU locations.	A NetNearU reseller wrote in to note that he resells the turnkey system either for $495 and gives venues 15 percent of the revenue in exchange for providing technical support and marketing, or for $695 with a 40 percent revenue cut for the location, but they're responsible for marketing and tech support themselves.
FatPort	FatPort Complete: A hands-off managed solution for $199 a month, including *Digital Subscriber Line* (DSL) service, technical support, and what they label FatPoint Express.	Up to 40 percent revenue share.
Joltage	Joltage Networks intends to use its proprietary software to foster what it calls *micro-ISPs* or hot spots. The software handles all the back-end details of running a WISP, including security, billing, and administration.	Joltage is following a revenue-sharing plan where it will split the $24.99 monthly access fee with the hot spot providers. In homage to Avon, Joltage will pay providers to bring others into the service. Joltage says micro-WISPs using its patent-pending software can opt to provide Internet access free of charge.

Source: 802.11b Networking News[8]

[8]"Turnkey Hotspots," a report from 802.11b Networking News, http://80211b.weblogger.com/turnkey_hotspots.html.

Reciprocal Compensation—Build It and They Will Pay You

One critical building block for the propagation of 802.11 is reciprocal compensation. Wi-Fi users must be able to access wireless networks wherever possible in order for this technology to take hold. One potential market driver for the building of public networks is reciprocal compensation. In reciprocal compensation, one operator pays another when a subscriber of the first operator logs into the network of the second operator.

A good example of this model can be found in the cell phone industry. Cell phone companies thrive on charging roaming fees. When a subscriber is making calls within his or her service area, the charge may be 10 cents a minute. However, once the subscriber travels outside his or her immediate service area and accesses another cell phone company's network, the subscriber is charged a roaming rate that may be as high as 80 cents or more per minute. The subscriber's cell phone company must share that revenue with the originating cell phone company.

When the Internet revolution took place in mainstream America in the mid-1990s, a number of *Internet service providers* (ISPs) became *Competitive Local Exchange Carriers* (CLECs) because under the legal regimen of the time, CLECs could command reciprocal compensation from *Incumbent Local Exchange Carriers* (ILECs) for dial-up Internet access calls. Thus, the CLEC boom was born. The prospect of Wi-Fi subscribers driving reciprocal compensation for network operators might be a major catalyst for the deployment of even more 802.11 APs, thus spreading public networks nationwide. The establishment of reciprocal compensation agreements between service providers and the creation and operation of clearinghouses where revenues for accessing various networks are distributed are well under way. Programs for reciprocal compensation, roaming fees, and revenue sharing may very well promote a build-it-and-make-money catalyst for the construction of public 802.11 access networks.

Turnkey operators are probably the key to the immediate future for the propagation of 802.11 as a residential broadband Internet technology. Currently, the technology is hobbled by a conservative mentality that focuses on a service footprint with a maximum radius of 50 meters—that is, it is confined to spaces where a significant number of subscribers are likely to congregate.

How then will the promise of true broadband Internet offering multiple megabits of bandwidth to the home be realized? The answer lies both in technology and business models. First, antenna technology is improving to where the range of an antenna can be measured in miles instead of meters. If the service can move out of the hot spot and into the neighborhood, broadband can become ubiquitous. Some new antenna products promise steerable technologies that will allow an antenna to offer service to a specific household miles from the antenna. With new antenna technologies, local operators could spring up to service their respective suburbs or small cities and towns.

Another technology hurdle that must be overcome is the bandwidth of the feed to the wireless AP. Currently, many hot spots can obtain speeds of only 1.54 Mbps from the local telephone company. As explored in Chapter 3, "Range Is Not an Issue," the deployment of a *wireless metro area network* (WMAN) would allow the distribution of wireless broadband citywide at speeds approaching 11 Mbps. Many hot spot operators resell bandwidth from a DSL or cable modem service, which is often less expensive than a T1.

The turnkey business model offers a number of advantages for both the local operator and the turnkey provider. The advantage for the local operator is the cost of maintaining a help desk. Other customer service functions are provided by the turnkey provider. Single calls to a help desk can cost a service provider tens of dollars per call. Furthermore, having network management provided by the turnkey provider also spares the local operator lots of money.

Making the Private Network Public—Layering

If public access hot spots can be established by enabling existing and proposed enterprise WLANs to deliver public Wi-Fi coverage, the market would be much closer to making wireless Wi-Fi availability ubiquitous. This approach is called *layering*. Layering uses existing WLANs from larger enterprises to provide the bandwidth for public networks (hot spots). Unlicensed enterprise systems often cover or spill coverage out onto prime urban property and locations frequented by potential Wi-Fi users.

Layering for enterprise networks involves installing a network management gateway that recognizes and differentiates between authorized users from the enterprise and public access visitors. Enterprise users can receive secure *IP Security* (IPSec) encrypted WLAN service—with their communications operating over a guaranteed bandwidth allocation securely separated from that of visitors making use of the network. Visitors are directed to a WISP login screen and experience typical public hot spot service. Layering provides additional revenue opportunities for WLAN implementations that can offset some of the implementation cost issues and hurdles faced by some IT managers. Figure 8-6 illustrates the process of layering.

Figure 8-6

Layering is the resell of corporate bandwidth to the public in nearby locations.

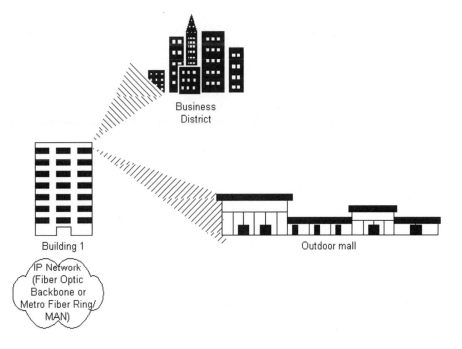

Business District

Building 1

IP Network (Fiber Optic Backbone or Metro Fiber Ring/ MAN)

Outdoor mall

Security and Bandwidth Management Security concerns are addressed by the gateway that manages the subnet of APs and assigns different roles to different groups or classes of users. One of these roles is designated for visitors who are blocked from accessing the enterprise systems and are channeled to and through the hot spot gateway software. The gateway can also perform wireless link encryption (supporting IPSec and *Point-to-Point Tunneling Protocol* [PPTP] encrypted tunnels) and authentication using a built-in database or links to existing *Remote Authentication Dial-In User Service* (RADIUS) servers. *Virtual LANs* (VLANs) are used to segment the traffic between private and public use. VLANs can also limit the bandwidth of the public use and cut back on it on an as-needed basis. This way, a business can sell its excess bandwidth to the public in a hot spot up to a mile away from the building, assuming it uses the right antenna technology.

The layering model vastly expands the community-based approach started by urban residential broadband users who began offering free Internet access over their DSL connections by letting others in their neighborhood connect through their 802.11b APs. Many existing and proposed enterprise WLAN locations could use layering to provide hot spot coverage. Enterprise offices that front onto public spaces and that frequently host visitors, auditors, or consultants can benefit both themselves and their tenants/visitors by offering WLAN layering. Location owners will also benefit from the improved enterprise WLAN implementation economics that will be realized by offsetting hot spot revenue sharing against capital and operating costs.[9]

WISP ROI for Wireless Access via 802.11

A strong motivation for the deployment of 802.11 access could come from incumbent ISPs. Most ISPs have long been dependent on coop-

---

[9]Tim Brooks, "Layering to Create Public Networks," www.wirelessreport.net/wirelessnetworks/july02/layeringtocreatepublichotspots.html, July 2002.

eration from incumbent telephone companies to deliver their services to their subscribers. Most telephone companies offer a competing ISP service. ISPs could free themselves of that dependency by deploying their own wireless access infrastructure. An ISP moving into this market space could endear itself with the small business community by offering a dial-a-bandwidth service in competition with the local telephone company. Bandwidth is usually sold by telephone companies in denominations of T1 (1.54 Mbps) and the rates range from $300 to $500 depending on the market and up to $10,000 per month depending on the distance. If an enterprise requires more than 1.54 Mbps, it is forced to buy another full 1.54 Mbps from its service provider at the going rate ($1,000 per month in many cases). Plans where the business is charged by the megabit or gigabit are also available. The company will also have to pay a monthly service fee to an ISP to connect to the Internet if that is not included in its bandwidth cost. Table 8-6 presents a comparison of these costs.

Offering Vo802.11 telephony services could also serve to distinguish the WISP from its competition and reduce churn. In addition, system integration services for the installation of WLAN equipment could be another source of revenue for the WISP.

Table 8-6

Cost comparison of services provided by a telephone company versus services available via a WISP

Component	Regional Bell Operating Company (RBOC)	WISP
Data	1.54 Mbps at $500 a month (includes Internet access)	1.54 Mbps, less than $100 a month.
Local phone service	$50 per line a month	No cost for interoffice calls on the WLAN to other IP addresses; calls to PSTN numbers are 3 cents per minute.
Long distance	3 to 7 cents per minute for all calls	Free interoffice calls on the WAN; other calls are 3 cents per minute.

Viral Growth of 802.11 Networks: Community Networks and the Mom 'n Pop WISP

Community Networks as an Industry Much has been written in the media about activists who are building community wireless networks. Wi-Fi may enable the creation of a new ISP industry that does not depend on local connections controlled by incumbent carriers such as the ILECs in the United States or the government *Post, Telegraph, and Telephones* (PTTs) elsewhere.

The scenario for community networks as an industry is likely to play out as it did for the mom 'n pop ISPs of the early Internet age (mid-1990s). However, just as small ISPs served as a goad for major carriers to provide Internet service, the grassroots wireless operators will do the same for wireless broadband. The grassroots operators will also likely serve as fertile technology and application incubators —their operators and constituents are just the kind of tech-savvy enthusiasts that produce the best and brightest ideas.

One outcome of the community wireless movement will be increased wireless broadband for less-wealthy areas and other forms of public service. If this happens, some of the promise of telecommunications as a democratizing force may be fulfilled. Particularly of

Table 8-7

Potential cost savings in household telecommunications costs using 802.11 WISP versus conventional service providers

Component	Conventional	WISP
Local phone service (two lines)	$50	$50 (assuming use of a cell phone)
Long distance	$100 ($.07 a minute)	$0 (assuming all calls are VoIP)
Video (cable versus video on demand)	$50	$0
Internet	$25	$0
Broadband service (DSL or cable)	$40	$45
Totals	$265	$95

interest will be the fate of the grassroots movement outside the United States, where microeconomic, community-owned approaches may be the best option to provide services to the population base. Strong precedents have been made for this tactic, including cellular build-out in certain markets. Interestingly, many sociopolitical overtones accompany this movement—in a way, this kind of movement is akin to the growth of broadcasting in its early years and implies the same sort of increased freedom of information. Over the next few years, some interesting stories will likely emerge from the community wireless movement.

Timing Factors   The community wireless movement has already shown real growth worldwide, with many organizations and individuals seeking to provide blanket coverage over a wide area or community. Projects have sprung up in Aspen, New York City, London, and many other locations. Another interesting factor is municipal and state support for Wi-Fi build-out as a source of broadband connectivity for areas not covered by traditional service providers. One challenge that will impede growth of the community wireless movement will be economics; until issues of who will pay are worked out, the business model will not be supportable. This is comparable to the early days of Internet connectivity, where early ISPs had to find ways to bill customers for service.[10]

The Economic Benefits of Ubiquitous Broadband with Public Networks

A wave of opportunity for wireless broadband applications is in the making. Most of it lies in the form of broadband deployment. In their April 2001 white paper "The $500 Billion Opportunity: The Potential Economic Benefit of Widespread Diffusion of Broadband Internet

---

[10]Chris Fine, "Watch Out for Wi-Fi," a white paper from Goldman Sachs, September 26, 2002.

Access," Robert Crandall and Charles Jackson point to an economic benefit of $500 billion per year for the American economy if broadband Internet access were to become as ubiquitous as landline phones.

The remainder of this chapter assumes that it is considerably cheaper (both in terms of hardware and lawyers) to deploy wireless broadband Internet to a residence than a similar service that depends on wiring (copper wire from the phone company or coax cable from the cable TV company). Both telephone wires and cable TV coax cable are accessible by almost 90 percent of American households. Even if they were not, the cost of copper wire, for example, is 15 cents per foot. The physical cost of connecting a home to the Internet in most residential applications is not that high. However, for a new market entrant, gaining the right-of-way from private landowners and public utilities to get to those households in most cases will not be possible without costly legal procedures. The legal costs of running wire or cable to a residence may not be offset by the revenue generated from subscription fees from that residence.

Using 802.11 as a means of access does not require legal dealings for rights-of-way and, compared to wired infrastructure, can be deployed much more quickly. Based on the fact that only 8 percent of U.S. households have broadband access via either telephone wires or cable TV coax cable, it will be assumed that these wired means of access are, for a variety of reasons, inadequate for achieving the same level of penetration in the market as telephone service. As evidenced by the efforts of CLECs to offer competitive residential telephone service using incumbent telephone poles and other incumbent-owned and -operated facilities, it is far easier to bypass PSTN facilities than to utilize them via legal means. 802.11 only requires an AP to be accessible by a residence. A wireless service provider only needs to install an AP and turn up the service. The remainder of this chapter explores the benefits of ubiquitous residential broadband Internet access, assuming the ease and economy of Wi-Fi is a catalyst for achieving the same levels of penetration for broadband Internet access as residential telephone service has today.

As the uses of broadband multiply, the value to subscribers rises far above the monthly subscription price. This is the *consumer surplus* from the innovation. Producers of new services that rely on the broadband (for example, i-mode-type services, Net2Phone, and so on), of products used in conjunction with broadband service (soft-switches, media gateways, IP phones, and residential gateways), and even of the broadband service itself also benefit from the greater diffusion of broadband. The *producer surplus* that is generated by sales is a real benefit to producers and, therefore, to the economy. Currently, no more than 8 percent of American households subscribe to a broadband service; only slightly more than 50 percent subscribe to an Internet service of any kind; and 94 percent subscribe to ordinary telephone service.[11] If broadband becomes ubiquitous, it would resemble current telephone service in its household penetration.

Producer Benefits

One of the reasons many IP backbone carriers went bankrupt is that they could not deliver bandwidth to a broad market. The bottleneck to the last mile remains the access controlled largely by telephone companies with their ubiquitous twisted-pair copper wire. Cable TV companies now service a majority of American homes.

Like Ethernet, Wi-Fi is not just about the technology itself—it is about what the technology empowers and the applications that result from the technology. Understanding the indirect effects of Wi-Fi is key to an investor's ability to profit from the technology. Figure 8-7 illustrates the Wi-Fi pull-through phenomenon on various segments of the PC and communications tech food chain. The following pages explain in more detail how Wi-Fi affects each technology segment discussed in the following sections. One example is the proliferation in Wi-Fi-enabled laptops. Intel will include 802.11a and 802.11b technology on their computer chips. Wi-Fi will probably become as common in computers as *universal serial bus* (USB) or Ethernet connections are now.

---

[11]The number of broadband subscribers (DSL and cable modems) was 7.3 million as of March 2001. See Crandall, Robert and Charles Jackson, "Failure of Free ISPs Triggers First-Ever Dip to 68.4 Million Online Users: Cable Modem Boom Continues as DSL Sign-ups Lag," *Telecommunications Report*s (April 2001): 1. The estimates for Internet and telephone service are from the authors' tabulations using the Current Population Survey for August 2000.

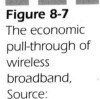

Figure 8-7

The economic
pull-through of
wireless
broadband,
Source:
Goldman Sachs

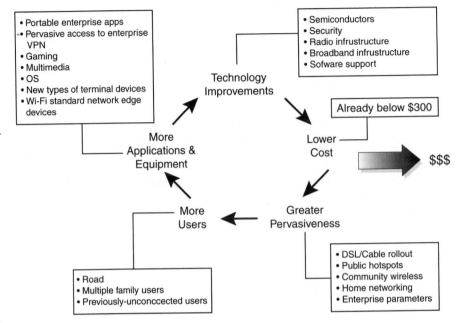

Wi-Fi will create a cycle of adoption that will drive technology purchases and upgrades by enterprises, retailers, service providers, government, and individuals for the three following reasons. First, it offers a means of delivery that is cheaper, simpler, smaller, and more convenient than a wired (telephone and cable TV) means of delivery. Wi-Fi service requires either the presence or installation of a broadband connection and/or traditional LAN equipment. The larger the environment—that is, the number of coverage areas, or hot spots, the number of users supported, and so on—the more infrastructure equipment and network bandwidth are required, thus spurring sales of APs, wireless routers, antennas, and so on.

Once Wi-Fi is available in a given area, it will spur the purchase of more mobile computers, *personal digital assistants* (PDAs), and other networked devices. This is particularly relevant in the home, where Wi-Fi enables broadband connections to be shared easily among multiple PCs and, ultimately, other devices. Major PC vendors are beginning to include Wi-Fi support now, and PC cards can be purchased for as low as $30, with prices continuing to drop. Chipmakers and laptop computer manufacturers now include Wi-Fi capabilities in their products.

New applications will drive the uptake of Wi-Fi. These market drivers include home networking, home multimedia, smart appliances, and voice over 802.11. These applications require new platforms such as home APs and Vo802.11 telephony devices.[12]

Computer Sales   The expansion of the demand for broadband will create additional demand for computers and networked home appliances. As of June 2000, approximately 40 percent of all U.S. households do not have a computer.[13] Clearly, these households are not equipped to connect to the Internet at any speed. Of the 60 percent of households with computers, many will need to upgrade their equipment to obtain greater processing speed, more *random access memory* (RAM), or greater hard-drive capacity. Still others will choose to buy more advanced equipment such as storage devices, MP3 players for music downloads, and *liquid crystal display* (LCD) projectors for viewing video downloaded via a high-speed Wi-Fi connection. The applications that the following sections explore (video and telephony) could very well drive much of the remaining 40 percent of households without computers to make the leap and install a computer in their homes.

Crandall and Jackson estimate broadband's stimulus on household purchases of broadband-related equipment would be that U.S. household spending on computer equipment, peripherals, and software would resume its growth rate of 14.3 percent per year during 1991 through 1995, rather than continuing at its growth rate of 10.4 percent per year during 1995 through 1999. If growth returns to its pace in 1991 to 1995, by 2006, total spending would be $80 billion, rather than $66 billion—an increase of $14 billion. By 2011, the difference would be $53 billion per year. If the broadband revolution

---

[12]Chris Fine, "Watch Out for Wi-Fi," a white paper from Goldman Sachs, September 26, 2002.

[13]The most recent estimate from the Bureau of the Census for June 2000 was 41.5 percent. More recent estimates from TNS suggest that about 50 percent of households now have access to the Internet. See TNS Telecoms, ReQuest Market Monitor National Consumer Survey, v.3 (2001).

accelerated household equipment expenditures by another 3 percent per year to 17.3 percent annual growth, the additional spending in 10 years would be $110 billion per year.[14]

Consumer Benefits

The most straightforward estimate of the value of the enhanced availability of broadband derives from information on consumer subscriptions to broadband services. Currently, no more than 8 percent of households subscribe to a broadband service; only slightly more than 50 percent subscribe to an Internet service of any kind; and 94 percent subscribe to ordinary telephone service. Were broadband to become ubiquitous, it would resemble current telephone service in its household penetration.

An Estimate Based on Current Demand   The price elasticity of demand is a relationship of the change in demand to the change in price. Given a current broadband penetration of 8 percent and an average price of the service of $40 per month, total broadband revenues may be estimated at $480 times 8.4 million, or $4 billion per year. Assuming that the demand for such service is linear with an elasticity of -1.0, the value of the service to these consumers—the consumer surplus—is $2 billion per year in addition to the $4 billion they pay. If the demand elasticity is -1.5, the consumer surplus falls to $1.4 billion.

If broadband spread to 50 percent of households at $40 per month through a shift of a linear demand curve with constant slope, the annual expenditure on the service would rise to $31.2 billion. At 50 percent penetration, the additional value to consumers would rise to between $80 billion and $121 billion per year at these two price elasticities. If broadband service became truly ubiquitous, similar to ordinary telephone service, annual consumer expenditures on the service would rise $58.7 billion per year, assuming the continued

---

[14]Robert Crandall and Charles Jackson, "The $500 Billion Opportunity: The Potential Economic Benefit of Widespread Diffusion of Broadband Internet Access," a white paper from Criterion Economics, LLC, July 2001.

shift of the linear demand curve had a constant slope and an annual price of $480. The additional value to consumers—over and above their expenditures on the service—would be $284 billion to $427 billion per year, assuming that the linear demand curve with a current elasticity of -1.0 or -1.5 simply shifted outward. Table 8-8 illustrates this price elasticity of demand.

Home Entertainment By the mid-1960s, a majority of homes in the United States had a television. This would now be called a wireless residential video service. Subscribers were limited in content to three channels of programming focused on the evening hours known as *prime time*. Those subscribers were forced to be present in front of their video monitors at precisely the time of the broadcast. There was no means of storing the program for viewing at a later time. The coming of cable TV and *videocassette recorders* (VCRs) in the following decades added some flexibility to the TV viewing experience.

Before cable TV and VCRs, subscribers were entirely at the mercy of the programmers. They had to watch what the programmers offered. The ability to choose programming drove the growth of cable and VCRs, leading to myriad new businesses, including cable TV companies and video rental firms. For most programming (films and prime-time TV shows), the production costs were very high and distribution was costly. This presented a high barrier to entry for any competitors.

Table 8-8

Estimated ultimate annual consumer surplus from increased broadband penetration

Current Price Elasticity of Demand ($ Billions)		
	−1.5	−1.0
At 8 percent penetration	1.4	2.0
At 50 percent penetration	80.0	121.0
At 94 percent penetration	284.0	427.0

Source: Crandall and Jackson

The coming of the Internet held out some promise for a break in this regime. However, one barrier remained to a mass market for video delivered via the Internet—the bandwidth bottleneck. At the time of this writing, less than 10 percent of U.S. households enjoy broadband Internet. Dial-up Internet access at speeds up to 56 Kbps is too slow to download a feature-length film. DSL Internet access with speeds of around 256 Kbps is markedly better, but still requires a few hours of download time to receive a feature-length film.

What exactly constitutes a viable online video service? Is a short news or sports broadcast from a TV station made available via the Web a viable video service paid or unpaid? Is delivering an advertorial from an advertiser a commercially viable video service? Does the video have to be viewed in real time using video streaming or can the video file(s) be scheduled for an overnight download on the subscriber's computer? Can the subscriber have multiple computers to download video files around the clock? Is a feature-length film the only commercially viable form of Internet-based video service? Or are 30-minute serials or sitcoms also commercially viable online? What if the subscriber downloads or shares video files of all shapes and sizes at a completely random sequence, stores them in a digital video library, and views them completely at his or her leisure. One efficient form of delivery is to cache popular video files (films) on a local server, thus speeding the download. The subscriber can then use a PC to download video for viewing at his or her leisure, not unlike Personal Video Recorder technology like DirectTV's TiVo.

The freedom to obtain video files at little or no expense and store them in personal libraries presents a marked departure from the programmer-centric paradigm of the 1960s. This marks a shift in power of selection from the content programmer to the subscriber as the subscriber now chooses what he or she will view. More importantly, it determines what the subscriber is willing to pay for—the ultimate test of a product's commercial viability in a free market economy.

At the time of this writing, file sharing of video files, including feature-length films, has been available online for years. In October 2002, a start-up firm named Cflix launched a paid video download service offering a variety of feature-length films and some video serials such as the popular animation *South Park*. One month later, a consortium of Hollywood firms launched a service called MovieLink, which offers recent Hollywood releases for a fee per download via broadband Internet connections.

Broadband Internet via 802.11 with download speeds at around 6 Mbps makes the delivery of video services much more viable. A number of web sites coordinate the sharing of even the most recent releases of feature-length films. This sharing of files validated the technical viability of video file sharing and downloading long before the emergence of Cflix and MovieLink.

Making Money with Online Video   So how does one make money on video on the Internet? MovieLink charges $3 to $5 per feature-length film. The subscriber doesn't get to keep it for that price. After downloading, the buyer has 30 days to activate the movie, but once the video file is opened, the subscriber can watch it as often as he or she wants within a 24-hour period. Then it disappears from the hard drive.

A Cflix subscription, which includes some basic programming, costs $4 per month—considerably cheaper than other video services. Cflix subscribers pay an additional $1.99 for older movies and $3.99 for new releases. They can attach equipment to their computers that enables them to watch the movies on a TV set.[15]

MovieLink is backed by Warner Brothers, Paramount, Universal Studios, MGM, and Sony Pictures, and offers people with broadband Internet connections downloads of about 175 recent and classic movies. To be truly useful, MovieLink has to offer thousands of titles, not a couple hundred. Even with an Internet connection capable of transferring 1 Mbps, it would take around 10 hours to download a 2-hour movie at that compression setting. More advanced compression technology, such as MPEG-4, may be able to cut the bandwidth in half while maintaining equal quality, but the download times would still be prohibitive.[16]

---

[15]Dan Luzadder, "Video Service Gives It That New College Try," *Denver Post*, www.denverpost.com/cda/article/print/0,1674,36%257E33%257E940472,00.html, October 22, 2002.

[16]Steven Wildstrom, "Hardly a Boffo Start for MovieLink," *Business Week*, www.businessweek.com/technology/content/nov2002/tc20021115_6283.htm, November 15, 2002.

How can the economic scope and scale of video over broadband be measured? History is probably the best teacher. The multichannel video revolution of the 1980s and 1990s created enormous value for consumers. This explosion in choice created between $77 billion and $142 billion in annual value beyond the costs of the service (consumer surplus).

The Napster sensation probably provides only a prelude to what is possible over household broadband connections. At its height, Napster traffic was over one-third of the total traffic on the Internet. Downloading motion pictures or other video material, interactive television, interactive games, and even home editing of digitized entertainment material have not even begun in earnest. It is reasonable to assume that eventually the contribution to the consumer surplus of new video and other entertainment options created by the widespread diffusion of broadband Internet access would be at least as great as that already created by cable TV and direct broadcast satellites. This provides the model with an estimate of $77 billion to $142 billion per year. This estimate is consistent with the remarkable growth in home-entertainment spending by U.S. consumers since 1980, the year that cable began to grow as the result of the FCC's deregulation of cable signal carriage (see Table 8-9). Total expenditures have risen by about $56 billion since 1980 (nominal dollars). If the price elasticity of demand at the current prices is substantially less than one, the consumer surplus from this increase could easily be more than $100 billion per year. Crandall and Jackson forecast the next broadband revolution—delivered by the broadband Internet (potentially wireless)—should be of equivalent value.[17]

Broadband Access and Telephone Services

A previous chapter outlined the technical aspects of routing voice over 802.11. What are the economics of replacing dial tone provided by the telephone company with voice over 802.11? As Vo802.11 is

---

[17]Robert Crandall and Charles Jackson, "The $500 Billion Opportunity: The Potential Economic Benefit of Widespread Diffusion of Broadband Internet Access," a white paper from Criterion Economics, LLC, July 2001.

Table 8-9

Total spending
on home
entertainment
(1980–1999)

Year	Expenditures by U.S. Households ($ Millions)	Year	Expenditures by U.S. Households ($ Millions)
1980	4,657	1990	29,822
1981	6,292	1991	32,160
1982	8,199	1992	34,009
1983	10,374	1993	38,016
1984	12,742	1994	39,513
1985	14,708	1995	42,380
1986	16,915	1996	46,647
1987	19,869	1997	50,730
1988	23,250	1998	55,231
1989	23,685	1999	60,765

VoIP using 802.11 as a means of access, the best approach is to study how VoIP replaces the PSTN dial tone in residential and small business telephone service.

Advantages of Vo802.11   Vo802.11 has a number of distinct economic advantages over using the PSTN and cell phone services. First, the cost of cell phone service is decreased by using Vo802.11 telephony in an office or any 802.11-serviced locale. Some new technologies allow a voice-enabled PDA to be dual-channel 802.11 and *Code Division Multiple Access/Global System for Mobile Communications* (CDMA/GSM). This capability enables an employee to talk over the 802.11 WLAN at the office or 802.11-serviced home or home office when in those offices. Once the employee leaves the office, he or she can switch over to a cell phone service provider if he or she has to make or receive calls.

Using Vo802.11 in the office can eliminate the cost of long-distance interoffice phone bills. Nearly 70 percent of corporate telephony is interoffice calling. This is an expense that can be eliminated by moving a company's telephony onto its corporate network. If 802.11

becomes a primary means of access within the company, then Vo802.11 would potentially eliminate much of a firm's phone bill.

A firm could eliminate all of its interoffice long-distance expenses by deploying a VoIP and 802.11 system. Calls routed over the corporate WAN would free the company from costs associated with long-distance phone service. Local phone service costs could be eliminated as well. If firms employed dual-frequency telephone handsets, all interoffice calls could be made on the corporate WAN. Local calls could also be routed to other 802.11 or IP-enabled handsets without contact with the PSTN. Other handsets could be reached using the cell phone network.

Soon the demand for broadband will reflect not only the growing potential uses of the Internet, but also the prospect for using these broadband connections to obtain voice telephone services currently provided over a narrowband connection. The use of broadband access to carry voice—ordinary telephone calls—as well as data will deliver substantial savings to consumers that are not captured in current demand estimates. Voice communications can be compressed, put in packets, and sent over an IP connection.

Microsoft Windows XP includes *Session Initiation Protocol* (SIP) software, which enables voice calls over the Internet—usually all one has to do to enable voice communications on a networked PC is to plug a headset into the audio ports on a computer and run the software. A broadband connection can support several voice connections —the exact number depends on the speed of the connection and the degree of compression of the voice signal. The current structure with two networks in the home (a voice and an IP network) and two connections to the outside world (a narrowband analog connection and a high-speed digital connection) appears inefficient. However, the transition to Internet telephony will take many years.

The cost savings from integrated access will be significant. Reliable Internet telephony would eliminate the need for second or third lines in households for teenagers or fax machines. The FCC estimated that the average household spent $55 per month on local and long-distance telephone service in 1999, and each household with telephone service had 0.289 additional lines.[18]

---

[18]FCC, "Trends in Telephone Service," 2nd Report, www.fcc.gov/Bureaus/ Common_Carrier/Reports/FCC-State_Link/IAD/trend200.pdf, 2000.

Within a few years, broadband access will permit consumers to substitute other services for these services that now cost $55 per month. The FCC estimates that the average residence spends $34 per month for local telephone service and $21 for long-distance telephone service. Part of that local telephone service cost is for the loop that is used for the broadband service. Consumers continue to incur most of those loop costs when broadband service is used, but they avoid the cost of the analog line card, the voice switch, and the voice transmission lines. Vo802.11 should lower the costs of both local and long-distance telephone service while providing residences with the equivalent of several telephone lines. Crandall and Jackson estimate that such savings could average $25 per month per household. In addition, households with broadband service would get the equivalent of multiple voice telephone lines. It is estimated that this additional service or option of service could be worth $10 per month to the average household.[19] Thus, in the longer run (say, a decade from now), broadband access could deliver voice communications benefits of about $35 per month, or $420 per year, to the average household with telephone service. If it is assumed that 122.2 million households have telephone service, these benefits would total $51.4 billion per year, assuming no growth in voice usage occurs. The actual value could be much higher.

The substantial economic benefits (principally savings from expenditures on telephone service) created by providing multiple services over a high-speed line almost cover the cost of a high-speed line—it has been estimated that benefits of $35 per month are created by a broadband connection that costs $40 per month. These sav-

---

[19]The FCC's numbers indicate that the average household with telephone service has 1.289 access lines and pays local service fees of $34 per month. Assuming that all lines cost the same (which is not quite right but is reasonable), the average household with telephone service in 1999 paid $7.62 per month for additional line service. If those households without a second line today place an average value of no more than $3.40 per month for a second line of service, then the average household will value a second line at $10 per month or more.

[20]This was calculated using a discount rate of 10 percent and assuming a 2 percent per year growth in the economy.

[21]These present values are 2.8 and 4.2 times the ultimate value of broadband adoption when evaluated at an interest rate of 10 percent per year.

ings are one reason why it seems reasonable to expect that the fraction of households with high-speed access services will ultimately approach the fraction that has telephone service today.

Speeding up the adoption of broadband access provides benefits sooner. The present value of the difference between the base adoption scenario and the much faster adoption scenario of the previous example is 140 percent of one year's worth of the benefits of ubiquitous broadband adoption by households.[20] Thus, if one assumed that, when fully adopted, broadband would generate benefits of $300 billion per year to U.S. consumers, a policy change that moves our society from the baseline adoption curve to the much faster curve would generate benefits with an NPV of about $420 billion.[21] The increase in the present value of producers' surplus would be about $80 billion. This acceleration is therefore worth $500 billion to U.S. consumers and producers. Table 8-10 lists the consumer benefits of universal broadband deployment.

How could speeding up the adoption of a technology have such massive benefits? The key lies in the substantial benefits that ubiquitous broadband can convey to consumers. Once virtually everyone has the service, the network effects from developing new services will become very large. Moving these benefits forward a few years can create very large benefits—even when evaluated from today's perspective. The powerful advantage of 802.11 over the current, dominant broadband technologies (DSL and cable modem) is that in the words of Clayton Christensen,[22] it is "cheaper, simpler, smaller, and more convenient to use." The lack of a requirement for wires and their incumbent, expensive rights-of-way could potentially give the wireless service provider a significant advantage over the wired incumbent.

---

[22]Christensen, Clayton, "Innovator's Dilemma—When New Technologies Cause Great Firms to Fail (Management of Innovation and Change Series)", Harvard Business School Press, 1997. This book coined the term "disruptive technology" which is defined as being "cheaper, simpler, smaller and more convenient to use."

Table 8-10

Summary of consumer benefits from universal broadband deployment ($ billions per Year)

Source	Low Estimate	High Estimate
Direct Estimates		
Broadband access subscription	284	427
Household computer and network equipment	13	33
Total Benefits	**297**	**460**
Alternative Estimates—Benefits Deriving from:		
Shopping	74	257
Entertainment	77	142
Commuting	30	30
Telephone services	51	51
Telemedicine	40	40
Total Benefits	**272**	**520**

Conclusion

The economics of the deployment of enterprise 802.11 networks may not be as obvious as a comparison of the installation and equipment costs of wired versus wireless. However, the advantages in employee efficiency justify the technology. There will be considerable debate over the next few years as to what model will propagate wireless broadband networks the soonest: a top-down model as driven by large multinational service providers or a bottom-up model driven by viral mom 'n pop service providers supported in a franchise-like distribution model. It is in the best national interest to deploy wireless broadband as quickly as possible in order for the U.S. economy (and any other national economy) to reap a $500 billion annual benefit.

Regulatory Aspects of 802.11

An objection often raised about 802.11 applications is that because the spectrum used by 802.11 (2.4 GHz for 802.11b and 5.8 GHz for 802.11a) is unlicensed, it will inevitably become overused (called *tragedy of the commons*) to a point of being unusable at which time the government (U.S. or other) will step in to control the spectrum, making it no longer free, thus costing the service provider his or her profit margin and relegating the market to deep-pocketed monopolists.

This chapter explores the considerations wireless service providers should take into account when deploying service on unlicensed 802.11 bands. Next, it explores a new initiative from the *Federal Communications Commission* (FCC), which heralds a change to spectrum management that may actually serve to liberalize the FCC's approach to what spectrum is unlicensed. Finally, the chapter covers an initiative in the U.S. Congress to free more spectrum for the use of broadband wireless Internet applications. If anything, it appears that the U.S. government is developing a policy to encourage the use of unlicensed spectrum.

The Current Regulatory Environment

Even though 802.11 operates in unlicensed spectrum, a service provider must know a number of things in order to stay out of trouble with state and federal authorities. The following paragraphs outline the most prominent problem areas.

Spectrum is managed by a number of different organizations. The most visible to the general public is the FCC. The FCC manages civilian, state, and local government usage of the radio spectrum. The FCC regulations are contained in the "Code of Federal Regulations, Title 47."

At the time of this writing, the FCC has very limited resources for enforcement, as the trend for the last couple of decades has been deregulation and the reduction of staffing in the enforcement bureaus. The *National Telecommunications and Information Administration*

(NTIA), which works with the *Interdepartmental Radio Advisory Committee* (IRAC), also manages federal use of the spectrum.

The following sections are a brief overview of what a service provider needs to be concerned about when operating in unlicensed spectrum. This synopsis was provided by Tim Pozar of the Bay Area Wireless Users Group based on many years experience advising friends and clients on what they can and cannot do with unlicensed spectrum.

Power Limits

Ideally, a well-engineered path has just the amount of power required to get from point A to point B with good reliability. Good engineering limits the signal to only the area being served. This has the effect of reducing interference and enabling more efficient use of the spectrum. Using too much power covers more area than is needed and could potentially interfere with other users of the band. Because 802.11 is designed for short-range use, such as offices and homes, it is limited to very low power.

802.11b—Its Relationship to FCC Part 15, Section 247   Regulatory aspects of 802.11 per Part 15 are oriented around the applications of the wireless technologies. The distinction is point-to-multipoint and point-to-point.

Point to Multipoint   802.11 service providers are allowed up to 30 dBm or 1 watt of *Transmitter Power Output* (TPO) with a 6 dBi antenna or 36 dBm or 4 watts *effective isotropic radiated power* (EIRP). The TPO needs to be reduced 1 dB for every decibel of antenna gain over 6 dBi.

Point to Point   The FCC encourages the use of directional antennas to minimize interference to other users. In fact, the FCC is more lenient with point-to-point links by requiring only the TPO to be reduced by one-third of a decibel instead of a full decibel for point to multipoint. More specifically, for every 3 dB of antenna gain over a 6 dBi antenna, a *wireless Internet service provider* (WISP) needs to reduce the TPO 1 dB below 1 watt. For example, say a 24 dBi antenna is 18 dB over a 6 dBi antenna. This requires lowering a 1 watt (30 dBm) transmitter 18/3 or 6 dB to 24 dBm or 1/4 watt.

802.11a—FCC Part 15, Section 407 Regulatory aspects of 802.11 per Part 15 are oriented around the applications of the wireless technologies. The distinction is point-to-multi-point and point-to-point.

Point to Multipoint As described before, the *Unlicensed National Information Infrastructure* (U-NII) band is chopped into three sections. The low band runs from 5.15 to 5.25 GHz, with a maximum power of 50 mW (TPO). This band is meant to be in-building only as defined by the FCC's Rules and Regulations Part 15.407(d) and (e):

(d) Any U-NII device that operates in the 5.15–5.25 GHz band shall use a transmitting antenna that is an integral part of the device.
(e) Within the 5.15–5.25 GHz band, U-NII devices will be restricted to indoor operations to reduce any potential for harmful interference to co-channel MSS operations.

The middle band runs from 5.25 to 5.35 GHz, with a maximum power limit of 250 mW. Finally, the high band runs from 5.725 to 5.825 GHz, with a maximum transmitter power of 1 watt and antenna gain of 6 dBi or 36 dBm or 4 watts EIRP.

Point to Point As with 802.11b, the FCC does give some latitude to point-to-point links in Part 15.407(a)(3). For the 5.725–5.825 GHz band, the FCC allows a TPO of 1 watt and up to a 23 dBi gain antenna without reducing the TPO 1 dB for every 1 dB of gain over 23 dBi.

Part 15.247(b)(3)(ii) does allow the use of any gain antenna for point-to-point operations without having to reduce the TPO for the 5.725–5.825 GHz band.

Interference

Of course, interference is typically the state of the signal you are interested in while it is being destructively overpowered by a signal you are not interested in. The FCC provides a specific definition of *harmful interference* in Part 15.3(m):

> (m) Any emission, radiation, or induction that endangers the functioning of a radio navigation service or of other safety services or seriously degrades, obstructs, or repeatedly interrupts a radiocommunications service operating in accordance with this chapter.

Because this band has other users, interference will be a factor in 802.11 deployments. The 2.4 GHz band is a bit more congested than the 5.8 GHz band, but both have their co-users. Table 9-1 describes the other users of this spectrum and what interference mitigation may be possible for each.

Devices that Fall into Part 15   This band includes unlicensed telecommunications devices like cordless phones, home spy cameras, and *frequency-hopping spread spectrum* (FHSS) and *direct sequence spread spectrum* (DSSS) *local area network* (LAN) transceivers. Operators have no priority over or parity with any of these users, and any device that falls into Part 15 must not cause harmful interference to all licensed and legally operating Part 15 users and must accept interference from all licensed and legally operating Part 15 users. This is stated in Part 15.5(b) and (c):

> (b) Operation of an intentional, unintentional, or incidental radiator is subject to the conditions that no harmful interference is caused and that interference must be accepted that may be caused by the operation of an authorized radio station, by another intentional or unintentional radiator, by ISM equipment, or by an incidental radiator.
> (c) The operator of a radio frequency device shall be required to cease operating the device upon notification by a Commission

Table 9-1

Spectrum
allocation for
802.11 and
co-users

Part/User	Start GHz	End GHz
Part 87	0.4700	10.5000
Part 97	2.3900	2.4500
Part 15	2.4000	2.4830
Fusion lighting	2.4000	2.4835
Part 18	2.4000	2.5000
Part 80	2.4000	9.6000
Industrial, Scientific, and Medical (ISM)—802.11b	2.4010	2.4730
Part 74	2.4500	2.4835
Part 101	2.4500	2.5000
Part 90	2.4500	2.8350
Part 25	5.0910	5.2500
U-NII low	5.1500	5.2500
U-NII middle	5.2500	5.3500
Part 97	5.6500	5.9250
U-NII high	5.7250	5.8250
ISM	5.7250	5.8500
Part 18	5.7250	5.8750

Source: Tim Pozar, Bay Area Wireless Users Group from FCC sources

representative that the device is causing harmful interference. Operation shall not resume until the condition causing the harmful interference has been corrected.

Operators of other licensed and nonlicensed devices can inform you of interference and require that you terminate operation. It doesn't have to be a Commission representative.

Using 802.11b, you can interfere even if you are on different channels, as the channels are 22 MHz wide and only spaced 5 MHz apart.

Channels 1, 6, and 11 are the only channels that do not interfere with each other. (See Table 9-2.)

Devices that Fall into the U-NII Band   Unlike the 2.4 GHz band, this band does not have overlapping channels. The lower U-NII band has eight 20 MHz wide channels. You can use any of the channels without interfering with other radios on other channels that are within earshot. Ideally, it would be good to know what other Part 15 users are out there. Looking into groups under the banner of Freenetworks is a good place to start.

ISM—Part 18   This band is also an unlicensed service. Typical ISM applications are the production of physical, biological, or chemical effects such as heating, the ionization of gases, mechanical vibrations, hair removal, and the acceleration of charged particles. This band carries ultrasonic devices such as jewelry cleaners and ultrasonic humidifiers, microwave ovens, medical devices such as

Table 9-2

Spectrum bands of 802.11b

Channel	Bottom (GHz)	Center (GHz)	Top (GHz)
1	2.401	2.412	2.423
2	2.406	2.417	2.428
3	2.411	2.422	2.433
4	2.416	2.427	2.438
5	2.421	2.432	2.443
6	2.426	2.437	2.448
7	2.431	2.442	2.453
8	2.436	2.447	2.458
9	2.441	2.452	2.463
10	2.446	2.457	2.468
11	2.451	2.462	2.473

Source: Tim Pozar, Bay Area Wireless Users Group from FCC sources

diathermy equipment and *magnetic resonance imaging* (MRI) equipment, and industrial uses such as paint dryers (Part 18.107). *Radio frequency* (RF) should be contained within the devices, but other users must accept interference from these devices. Part 18 frequencies that could affect 802.11 devices are in the 2.400 to 2.500 GHz and 5.725 to 5.875 GHz ranges. As Part 18 devices are unlicensed and operators are likely clueless of the impact, it will be difficult to coordinate with them. Fusion lighting is also covered by Part 18.

Satellite Communications—Part 25 This part of the FCC's rules is used for the uplink or downlink of data, video, and so on to/from satellites in Earth's orbit. One band that overlaps the U-NII band is reserved for Earth-to-space communications at 5.091 to 5.25 GHz. Within this spectrum, 5.091 to 5.150 GHz is also allocated to the fixed-satellite service (Earth to space) for nongeostationary satellites on a primary basis. The FCC is trying to decommission this band for feeder use to satellites as "after 01 January 2010, the fixed-satellite service will become secondary to the aeronautical radio-navigation service." A note in Part 2.106, §5.446 also allocates 5.150 to 5.216 GHz for a similar use, except it is for space-to-Earth communications. There is a higher chance of interfering with these installations, as Earth stations deal with very low signal levels from distance satellites.

Broadcast Auxiliary—Part 74 Normally, the traffic on this band is *Electronic News Gathering* (ENG) video links going back to studios or television transmitters. These remote vehicles such as helicopters and trucks need to be licensed. Only Part 74 eligibles such as TV stations, networks, and so on can hold these licenses (Part 74.600). Typically, these transmitters are scattered all around an area, as TV remote trucks can go anywhere. This can cause interference to 802.11 gear such as *access points* (APs) deployed with omnidirectional antennas servicing an area. Also the receive points for ENG are often mountain tops and towers. Depending how 802.11 transmitters are deployed at these same locations, they could cause interference to these links. Wireless providers should consider contacting a local frequency coordinator for Part 74 frequencies that would be affected. There have been reports of FHSS devices interfering with these transmissions as the dwell time for this FHSS tends to punch holes in the video links. DHSS is less likely to cause interference to ENG users, but their links can cause problems with another person's 802.11 deployment. ENG frequencies that overlap 802.11 devices are 2.450 to 2.467 GHz (channel A08) and 2.467 to 2.4835 GHz (channel A09) (Part 74.602).

Land Mobile Radio Services—Part 90   For subpart C of this part, users can be anyone engaged in a commercial activity. They can use 2.450 to 2.835 GHz, but can only license 2.450 to 2.483 GHz. Users in subpart B would be local government. This would include organizations such as law enforcement, fire departments, and so on. Some uses may be video downlinks for flying platforms such as helicopters, also known as *terrestrial surveillance*. Depending on the commercial or government agency, coordination goes through different groups like the *Association of Public Safety Communications Officials* (APCO). Consider going to their conferences. You can also try to network with engineering companies that the government outsources to for their frequency coordination.

Amateur Radio—Part 97   Amateur radio frequencies that overlap 802.11b are 2.390 to 2.450 GHz and 5.650 to 5.925 GHz for 802.11a. They are primary from 2.402 to 2.417 GHz and secondary from 2.400 to 2.402 GHz. There is a *Notice of Proposed Rule Making* (NPRM) in with the FCC to change the 2.400 to 2.402 GHz range to primary. Amateurs are very protective about their spectrum.

Fixed Microwave Services—Part 101   Users of this band are known as *Local Television Transmission Service* (LTTS) and *Private Operational Fixed Point-to-Point Microwave Service* (POFS). This band is used to transport video. The channels are allocated from 2.450 to 2.500 GHz.

Federal Usage (NTIA/IRAC)   The federal government uses this band for radiolocation or radionavigation. Several warnings in the

FCC's Rules and Regulations disclose this fact. In the case of 802.11b, a note in the Part 15.247(h) gives the following warning:

> (h) Spread spectrum systems are sharing these bands on a non-interference basis with systems supporting critical government requirements that have been allocated the usage of these bands, secondary only to ISM equipment operated under the provisions of Part 18 of this chapter. Many of these government systems are airborne radiolocation systems that emit a high EIRP, which can cause interference to other users.

In the case of 802.11a, the FCC has a note in Part 15.407(a)(3) stating the following:

> The Commission strongly recommends that parties employing U-NII devices to provide critical communications services should determine if there are any nearby government radar systems that could affect their operation.

Laws on Antennas and Towers

In addition to concerns over the transmission of radio waves, the regulation of towers and antennas is a major concern for WISPs.

FCC Preemption of Local Law

The installation of antennas may run counter to local ordinances and homeowner agreements that would prevent installations. Thanks to the *Satellite Broadcasting and Communications Association* (SBCA), which lobbied the FCC, the FCC has stepped in and overruled these ordinances and agreements.

> So how does this apply to 802.11? This rule should only apply to broadcast signals such as TV, DBS, or MMDS. It could be argued that the provision for MMDS could cover wireless data deployment.

Height Limitations

The regulation that occurs most often in residential regulations is height restrictions on antennae. A number of regulatory agencies, both federal and local, have jurisdiction over antenna height.

Local Ordinances   Most, if not all, cities regulate the construction of towers. There is maximum height zoning of the antenna/tower (residential or commercial), construction, and aesthetic (for example, what color, how hidden, and so on) regulations.

Federal Aviation Administration (FAA) and the FCC Tower Registration   The FAA is very concerned about things that airplanes might run into. Part 17.7(a) of the FCC Rules and Regulations addresses this:

> (a) Any construction or alteration of more than 60.96 meters (200 feet) in height above ground level at its site.[1]

The FCC New Spectrum Policy

The American spectrum management regime is approximately 90 years old. In the opinion of FCC Chairman Michael Powell, it needs to be reexamined and taken in a new direction. Historically, four core assumptions have formed spectrum policy: (1) unregulated radio interference will lead to chaos; (2) spectrum is scarce; (3) government command and control of the scarce spectrum resource is the only way chaos can be avoided; and (4) the public interest centers on the government choosing the highest and best use of the spectrum. The following sections examine the four problem areas in spectrum management and provide their solutions.

---

[1]Tim Pozar, "Regulations Affecting 802.11 Deployment," a white paper from Bay Area Wireless Users Group, www.lns.com/papers/part15/, June 6, 2002.

Interference—The Problem

Since 1927, interference protection has always been at the core of federal regulators' spectrum mission. The Radio Act of 1927 empowered the Federal Radio Commission to address interference concerns. Although interference protection remains essential to the mission, interference rules that are too strict limit users' ability to offer new services; on the other hand, rules that are too lax may harm existing services. I believe the Commission should continuously examine whether there are market or technological solutions that can—in the long run—replace or supplement pure regulatory solutions to interference.

The FCC's current interference rules were typically developed based on the expected nature of a single service's technical characteristics in a given band. The rules for most services include limits on power and emissions from transmitters. Each time the old service needs to evolve with the demands of its users, the licensee has to come back to the Commission for relief from the original rules. This process is not only inefficient, but it can also stymie innovation.

Due to the complexity of interference issues and the RF environment, interference protection solutions may be largely technology driven. Interference is not solely caused by transmitters, which is the usual assumption on which the regulations are almost exclusively based. Instead, interference is often more a product of receivers; that is, receivers are too dumb, too sensitive, or too cheap to filter out unwanted signals. However, the FCC's decades-old rules have generally ignored receivers. Emerging communications technologies are becoming more tolerant of interference through sensory and adaptive capabilities in receivers. That is, receivers can sense what type of noise, interference, or other signals are operating on a given channel and then adapt so that they transmit on a clear channel that allows them to be heard.

Both the complexity of the interference task—and the remarkable ability of technology (rather than regulation) to respond to it—are most clearly demonstrated by the recent success of unlicensed operations. According to the Consumer Electronics Association, a complex variety of unlicensed devices is already in common use, including garage and car door openers, baby monitors, family radios, wireless headphones, and millions of wireless Internet access devices using Wi-Fi technologies. Yet despite the sheer volume of devices and their disparate uses, manufacturers have developed technology that allows receivers to sift through the noise to find the desired signal.

Interference—The Solution

The recommendation of the Interference Protection Working Group of the FCC's Spectrum Policy Task Force was that FCC should consider using the *interference temperature* metric as a means of quantifying and managing interference. As introduced in this report, interference temperature is a measure of the RF power available at a receiving antenna to be delivered to a receiver, that is, power generated by other emitters and noise sources. More specifically, it is the temperature equivalent of the RF power available at a receiving antenna per unit bandwidth, measured in units of °Kelvin (K). As conceptualized by the Working Group, the terms *interference temperature* and *antenna temperature* are synonymous. Interference temperature is a more descriptive term for interference management.

Interference temperature can be calculated as the power received by an antenna (watts) divided by the associated RF bandwidth (hertz) and *Boltzman's Constant* (equal to 1.3807 wattsec/°Kelvin). Alternatively, it can be calculated as the power flux density available at a receiving antenna (watts per meter squared), multiplied by the effective capture area of the antenna (meter squared), with this quantity divided by the associated RF bandwidth (hertz) and Boltzman's Constant. An interference temperature density could also be defined as the interference temperature per unit area, expressed in units of °Kelvin per meter squared and calculated as the interference temperature divided by the effective capture area of the receiving antenna (which is determined by the antenna gain and the received frequency). Interference temperature density could be measured for particular frequencies using a reference antenna with known gain. Thereafter, it could be treated as a signal propagation variable independent of receiving antenna characteristics.

As illustrated in Figure 9-1, interference temperature measurements could be taken at receiver locations throughout the service areas of protected communications systems, thus estimating the real-time conditions of the RF environment.

Like other representations of radio signals, instantaneous values of interference temperature vary with time and, thus, need to be treated statistically. The Working Group envisions that interference thermometers could continuously monitor particular frequency bands, measure and record interference temperature values, and compute the appropriate aggregate value(s). These real-time values could govern the operation of nearby RF emitters. Measurement devices could be designed with the option to include or exclude the on-channel energy contributions of particular signals with known characteristics such as the emissions of users in geographic areas and bands where spectrum is assigned to licensees for exclusive use.

The FCC could use the interference temperature metric to set the maximum acceptable levels of interference, thus establishing a worst-case environment in which a receiver would operate. Interference temperature thresholds could therefore be used, where appropriate, to define interference protection rights.

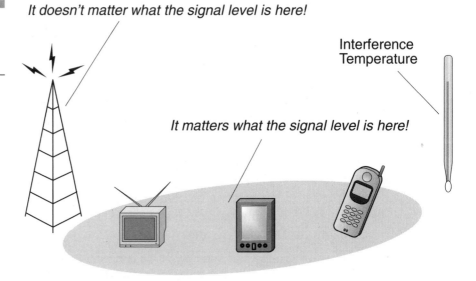

Figure 9-1
Interference temperature
Source: FCC

It doesn't matter what the signal level is here!

Interference Temperature

It matters what the signal level is here!

The time has come to consider an entirely new paradigm for interference protection. A more forward-looking approach requires that there be a clear quantitative application of what is acceptable interference for both license holders and the devices that can cause interference. Transmitters would be required to *ensure* that the interference level (or interference temperature) is not exceeded. Receivers would be required to *tolerate* an interference level.

Rather than simply saying your transmitter cannot exceed a certain power, the industry instead would utilize receiver standards and new technologies to ensure that communication occurs without interference, and that the spectrum resource is fully utilized. So, for example, perhaps services in rural areas could utilize higher power levels because the adjacent bands are less congested, therefore decreasing the need for interference protection.[2]

From a simplistic and physical standpoint, any transmission facility requires a transmitter, a medium for transmission, and a receiver. The focus on receiver characteristics has not been great in past spectrum use concerns; hence, a shift in focus is in order. The Working Group believes that receiver reception factors, including sensitivity, selectivity, and interference tolerance, need to play a prominent role in spectrum policy.[3]

Spectrum Scarcity—The Problem

Much of the Commission's spectrum policy was driven by the assumption that there is never enough for those who want it. Under this view, spectrum is so scarce that government forces rather than market forces must determine who gets to use the spectrum and for what. The spectrum scarcity argument shaped the Supreme Court's *Red Lion* decision, which gave the Commission broad discretion to

---

[2] Michael Powell, "Broadband Migration — New Directions in Wireless Policy," speech to Silicon Flatirons Conference, University of Colorado, Boulder, Colorado, October 30, 2002.

[3] Federal Communications Commission Spectrum Policy Task Force, "Report of the Interference Protection Working Group," November 15, 2002, 25.

regulate broadcast media on the premise that spectrum is a unique and scarce resource. Indeed, most assumptions that underlie the current spectrum model derive from traditional radio broadcasting and have nothing to do with wireless broadband Internet applications.

The Commission has recently conducted a series of tests to assess the actual spectrum congestion in certain locales. These tests, which were conducted by the Commission's Enforcement Bureau in cooperation with the Task Force, measured use of the spectrum at five major U.S. cities. The results showed that although some bands were heavily used, others either were not used or were used only part of the time. It appeared that these holes in bandwidth or time could be used to provide significant increases in communication capacity through the use of new technologies, without impacting current users. These results call into question the traditional assumptions about congestion. It appears that most spectrum is not in use most of the time.

Today's digital migration means that more data can be transmitted in less bandwidth. Not only is less bandwidth used, but innovative technologies like software-defined radio and adaptive transmitters can also bring additional spectrum into the pool of spectrum available for use.

Spectrum Scarcity—The Solution

While analyzing the current use of spectrum, the Task Force took a unique approach, looking for the first time at the entire spectrum, not just one band at a time. This review prompted a major insight—a substantial amount of white space is present that is not being used by anybody. The ramifications of this insight are significant. It suggests that although spectrum *scarcity* is a problem in some bands some of the time, spectrum *access* is a larger problem—how to get to and use those many areas of the spectrum that are either underutilized or not used at all.

One way the Commission can take advantage of this white space is by facilitating access in the time dimension. Since the beginning of spectrum policy, the government has parceled this resource in frequency and space. The FCC historically permitted use in a particular band over a particular geographic region often with an expectation of perpetual use. The FCC should also look at time as an additional dimension for spectrum policy. How well could society use this resource if FCC policies fostered access in frequency, space, and time?

Technology has, and now hopefully FCC policy will, facilitate access to spectrum in the time dimension that will lead to more efficient use of the spectrum resource. For example, a software-defined radio may allow licensees to dynamically rent certain spectrum bands when they are not in use by other licensees. Perhaps a mobile wireless service provider with software-defined phones will lease a local business's channels during the hours the business is closed. Similarly, sensory and adaptive devices may be able to find spectrum open space and utilize it until the licensee needs those rights for his or her own use. In a commercial context, secondary markets can provide a mechanism for licensees to create and provide opportunities for new services in distinct slices of time. By adding another meaningful dimension, spectrum policy can move closer to facilitating the consistent availability of spectrum and further diminish the scarcity rationale for intrusive government action.

Government Command and Control—The Problem

The theory back in the 1930s was that only the government could be trusted to manage this scarce resource and ensure that no one got too much of it. Unfortunately, spectrum policy is still predominantly a command and control process that requires government officials— instead of spectrum users—to determine the best use for spectrum and make value judgments about proposed and often overhyped uses and technologies. It is an entirely reactive and too easily politicized process.

In the last 20 years, two alternative models to command and control have developed, and both have flexibility at their core. First, the *exclusive use* or *quasi-property-rights model*, which provides exclusive, licensed rights to flexible-use frequencies, is subject only to limitations on harmful interference. These rights are freely transferable. Second, the *commons* or *open-access model*, which allows users to share frequencies on an unlicensed basis and has usage rights that are governed by technical standards, but has no right to protection from interference. The Commission has employed both models with significant success. Licensees in mobile wireless services have enjoyed quasi-property-right interests in their licensees and transformed the communications landscape as a result. In contrast, the unlicensed bands employ a commons model and have enjoyed tremendous success as hotbeds of innovation.

Government Command and Control of the Spectrum — The Solution

Historically, the Commission has had limited flexibility via command and control regulatory restrictions on which services licensees could provide and who could provide them. Any spectrum users who wanted to change the power of their transmitter, the nature of their service, or the size of an antenna had to come to the Commission to ask for permission, wait the corresponding period of time, and only then, if relief was granted, modify the service. Today's marketplace demands that the FCC provide license holders with greater flexibility to respond to consumer desires, market realities, and national needs without first having to ask for the FCC's permission. License holders should be granted the maximum flexibility to use, or allow others to use, the spectrum, within technical constraints, to provide any services demanded by the public. With this flexibility, service providers can be expected to move spectrum quickly to its highest and best use.

Public Interest—The Problem

The fourth and final element of traditional spectrum policy is the public interest standard. The phrase "public interest, convenience, or necessity" was a part of the Radio Act of 1927 and likely came from other utility regulation statutes. The standard was largely a response to the interference and scarcity concerns that were created in the absence of such a discretionary standard in the 1912 Act. The phrase "public interest, convenience, and necessity" became a standard by which to judge between competing applicants for a scarce resource and a tool for ensuring that interference did not occur. The public interest under the command and control model often decided which companies or government entities would have access to the spectrum resource. At that time, spectrum was not largely a consumer resource; it was accessed by a relatively select few. However, Congress wisely did not create a static public interest standard for spectrum allocation and management.

Public Interest—The Solution

The FCC should develop policies that avoid interference rules that are barriers to entry, that assume a particular proponent's business model or technology, and that take the place of marketplace or technical solutions. Such a policy must embody what has benefited the public in every other area of consumer goods and services—choice through competition and limited, but necessary, government intervention into the marketplace to protect such interests as access to people with disabilities, public health, safety, and welfare.

Congressional Moves Toward Spectrum Policy Reform

Contrary to the speculation that legislators would either not understand the potential contained in 802.11 technology or worse, seek to stifle its innovation for some unknown political objective, some members of Congress want to see Wi-Fi expand beyond wireless hot spots. Senators Barbara Boxer and George Allen announced plans to introduce broadband legislation in the 108th Congress that would require the FCC to make more spectrum available for Wi-Fi. The draft legislation calls for the FCC to allocate not less than 255 MHz of contiguous spectrum below 6 GHz for unlicensed use by wireless broadband devices.

The Boxer-Allen Bill also requires the FCC to develop guidelines for the expanded portion of spectrum allocated for Wi-Fi devices to avoid congestion. The senators said Wi-Fi is limited to a small portion of the spectrum, confining its development.[4]

Much of the current debate in Congress over broadband services has focused on two platforms—cable and DSL—and whether competition versus the deregulation of telecommunications is the best mechanism for encouraging broadband deployment. The bill was not yet completed at the time of this writing; however, a letter signed by Senators Allen and Boxer outlining the bill states, "This debate has reached an unproductive stalemate and fails to consider that other technologies are available that can jump-start consumer-driven investment and demand in broadband services."

The letter also explains the legislation is designed to "foster a third alternative mode of broadband communication, making more unlicensed spectrum available for exciting, new wireless technologies and requiring the FCC to design minimum rules of the road for broadband devices to operate in that spectrum."

Allen and Boxer claim the innovations and advances in the development of unlicensed wireless, radio-based networks (currently known as Wi-Fi) offer an additional means of delivering data at high speed and also enable the creation of new business models for delivering broadband connectivity.[5]

---

[4]Josh Long, "Senators Boxer and Allen to Introduce Broadband Legislation," www.phoneplusmag.com/hotnews/2bh2214134.html, November 22, 2002.

[5]Roy Mark, "Senators Aim to Wirelessly Jumpstart Broadband," http://siliconvalley.internet.com/news/article.php/154589, November 20, 2002.

Conclusion

This chapter outlined the current regulatory regime for 802.11 operators. It attacked the objection that not enough spectrum is available for a mass market deployment of 802.11 and that government interference in this area will only lead to a stultifying regulatory regime, which could kill 802.11 as a promising alternative as a last-mile solution to telephone and cable TV companies. Recent studies and pronouncements by the FCC and members of the U.S. Senate indicate support for reforming the spectrum policy in promoting the deployment of 802.11 and its related technologies.

Conclusion

Eventually, the Internet will replace the *Public Switched Telephone Network* (PSTN), but many technological changes will have to take place in order for this forecast to come to fruition. One major element will be the establishment of a faster connection to the Internet from the residence and small enterprise. In the foreseeable future, broadband Internet access will be limited if it is to be delivered by the duopoly of cable modems and *Digital Subscriber Lines* (DSLs).

802.11 and associated wireless technologies pose the greatest promise for a relatively rapid rollout of ubiquitous broadband Internet access. Clayton Christensen's definition of disruptive technology is that which is "cheaper, simpler, smaller, and more convenient to use." When compared, infrastructure-wise, to DSL (from the telephone company) and cable modems, 802.11 undoubtedly meets Christensen's definition.

Objections to 802.11

Four major objections have been made to the proposal to adopt 802.11 as a Last Mile solution, that is, access to a greater network, be it the Internet or the PSTN. Those objections are range, security, *quality of service* (QoS), and concern over interference from other Part 15 users in the unlicensed spectrum. In order to replace the PSTN's copper wire or cable TV's coax cable as Last Mile access, 802.11 as a technology must meet or exceed the performance of those forms of access. In addition, 802.11 has qualities of its own where its adoption will outweigh any disadvantages it has vis-à-vis copper wire or cable's coax.

Range as Limitation

One of the first tasks of this book has been to overcome the misperception that 802.11 is limited in range to approximately 100 meters. The range of 802.11 is a function of power and antenna technology. In order for 802.11 to achieve wide acceptance in residential and *small office/home office* (SOHO) markets, it will have to offer a range that makes it economical for the service provider to cover American suburbs. Achievements in power levels and directional antenna design have enabled 802.11 to facilitate point-to-point access in excess of 20 miles. Phased array antenna technology offers the potential of a "wireless switch" where different subscribers in a locale covered by one phased array antenna can receive differing levels of bandwidth and service.

The next question focuses on how bandwidth is delivered to a residential subscriber if the only source of bandwidth is the telephone company's T1 service. Developments in tiered network technologies and 802.16 offer the promise that bandwidth in excess of 100 Mbps can be distributed to subcarriers who can then deliver service to their subscribers and microcarriers at speeds of 11 Mbps (802.11b), 20 Mbps (802.11g), or 54 Mbps (802.11a). *Wireless metro area networks* (WMANs) have the potential to service large enterprise customers and subcarriers alike. The construction of infrastructure for this model of bandwidth distribution is far less expensive than fiber optic or other broadband schemes.

Security Fixes for 802.11

Another major objection to 802.11 is the misperception that it is inherently insecure. 802.11 contains a number of security mechanisms; the most prominent is *Wired Equivalent Privacy* (WEP). Any network security plan must use an equation that balances the value of information on the network (military intelligence, bank records, or email jokes) with the threat to the network (foreign intelligence agencies and inquisitive neighbors) with the cost of security measures in dollar terms (*virtual private networks* [VPNs], firewalls, and so on). 128-bit WEP contained in 802.11 is probably adequate for most residential and small enterprise applications. Users that need greater security can add VPNs, firewalls, and other additional measures to strengthen security on their network.

QoS or fear of interference from other Part 15 wireless users is perhaps the chief objection to 802.11. Major strides have been made in 802.11 QoS with the introduction of 802.11e, which is specifically aimed at improving QoS in 802.11 networks. 802.11e is based on over a decade of experience in the design of *wireless local area network* (WLAN) protocols and was built from the ground up for real-world wireless conditions. 802.11e is backwards compatible with 802.11; that is, non-802.11e terminals can receive QoS-enabled application streams.

QoS on 802.11 Networks

802.11 networks are potentially capable of delivering QoS comparable to the PSTN. It should be noted that the *Regional Bell Operating Companies* (RBOCs) have been losing phone lines to cell phone service providers at an alarming rate (for the RBOCs) over the last year. In fact, the RBOCs have recorded, percentage-wise, their first decline in lines in use since the Great Depression. Cell phone service is admittedly inferior in quality to that of the PSTN. The motivating factor, however, for landline customers to drop their service from the RBOC is the convenience offered by the cell phone as well as certain price advantages (free long distance in offpeak hours). The point here is that, ultimately, the QoS of the PSTN is not an absolute requirement for consumers. Consumers, in the case of cell phones, have traded off QoS for convenience and price. The PSTN is doomed if it must compete with 802.11 in that 802.11 using 802.11e potentially delivers a comparable QoS in both voice and data services while offering data rates up to 11 Mbps (compare with most DSL plans at 256 Kbps).

Voice over 802.11 (Vo802.11)

Segueing from QoS is the concept of routing voice traffic over 802.11 networks. The focus on QoS makes routing voice over 802.11 possible. Here is another performance parameter where 802.11 must match the capabilities of the PSTN in order for the technology to achieve wide acceptance. The Vo802.11 industry is very much in its infancy and is largely limited to in-building enterprise applications interfacing with conventional *private branch exchanges* (PBXs). However, some case studies point to the use of 802.11 to transmit voice point to point in a range of 19 miles. This is service that competes directly with the telephone company's Last Mile monopoly. The advent of softswitch technologies and *Internet Protocol* (IP) backbones makes it entirely possible to route phone calls that never touch the PSTN.

Economics of 802.11

Money makes the world go around. 802.11 offers a number of economic advantages over wired networks. In enterprise networks where 802.11 is only marginally cheaper to deploy than wired networks, 802.11 excels over wired network technology in that it offers the potential to make employees more efficient. It also reduces the costs of moves, adds, and changes. The chief market for 802.11 services at the time of this writing is in hotspots. Hotspots cover limited areas such as airports, coffee shops, and hotels. From these hotspots, 802.11 will gradually spread to a point where whole metropolitan areas are covered.

One significant economic advantage of 802.11 over the copper wire of the telephone company is the cost of deploying service. If a firm wants to compete with the phone company for service, it has two options. The first is to lease facilities from the phone company (called *unbundled network elements* [UNEs]), which are fairly economical if the competitor can get by the legal barriers the incumbents have been known to establish. The second is to deploy their own copper wire to the household, which can cost from $1,000 per household for an urban location to $10,000 for rural deployment. Considering the subscriber may only order basic phone service at $30 a month, the *return on investment* (ROI) on wired service is too long to be considered feasible; hence, only 8 percent of Americans have their choice of local telephone service providers.

The chief expense to deploy 802.11 service in a residential market is the *customer premise equipment* (CPE), which may run as much as $300 per residence and is falling quickly as new vendors enter the market. The expense of offering wireless service to a household is potentially considerably less than deploying new copper wire service. It is this difference in cost that will accelerate the deployment of wireless broadband service. The debate rages as to whether this roll-out will occur from the top down from existing service providers or from the bottom up from new market entrants. Perhaps the competition between the two camps will expedite the rollout of this service much to the benefit of the consumer. Economists Charles Jackson and Robert Crandall estimate that ubiquitous broadband would reap a $500 billion annual benefit to the U.S. economy.

Regulatory Environment for 802.11

Finally, another objecting view toward 802.11 is the fact that it uses unlicensed spectrum, which means any other competing service or appliance can introduce interference in the network, resulting in poor service to the end user. This is known as the tragedy of the commons. The alternative to unlicensed spectrum is licensed spectrum. Spectrum licenses for a particular frequency or band of frequencies are sold at auction. Licenses for spectrum dedicated to *third-generation* (3G) services offered by cell phone companies have committed those firms to billions of dollars of license costs. Given the cost of spectrum, few service providers can enter that market space. Other objections run the possibility that incumbents will lobby the government to make "free" spectrum no longer "free" any more.

The solution for the tragedy of the commons is twofold. First, it is entirely possible to engineer around interference regardless of the source. Secondly, the U.S. government is about to alter its spectrum management policy, which is based on the Radio Act of 1927. The *Federal Communications Commission* (FCC) has issued a series of reports that study issues such as spectrum scarcity, interference, government command and control of the spectrum, and the public's interest. It is possible that the FCC will offer sweeping reform of spectrum management that will favor *wireless Internet service providers* (WISPs).

Congress may also get into the act. At the time of this writing, new legislation is being introduced into Congress that would open 255 MHz of spectrum for unlicensed usage. With this much more spectrum to use, less likelihood of interference exists.

Potential Results of Ubiquitous Wireless Broadband Internet Service

What would be the benefits to society of ubiquitous wireless broadband service? The possibilities are limited only by the imagination. The following pages delve into looking at where 802.11 could benefit government and society.

National Defense Residential Broadband Network (NDRBN)

A number of new technologies have arrived on the market that make the deployment of broadband Internet access cheaper, simpler, and more convenient to deploy. Incumbent quasi-monopolistic service providers have failed to make broadband Internet access a priority for their customers and for national defense. In the interest of homeland security, it is imperative that our national leaders quickly adopt a policy for building the National Defense Residential Broadband Network.

The Problem

The federal government has been lambasted for failure to internally share intelligence that may have prevented or at least mitigated the attacks of September 11, 2001. Investigations into these allegations reveal outdated computer systems in the intelligence and security services at all levels of government. For a variety of reasons, intelligence and law enforcement agencies experienced difficulties in sharing information. Technologies developed in the dot-com boom could have overcome many of the technological shortfalls of these government networks. Relatively simple technologies such as restricted access web sites, encrypted email, video conferencing, and *voice over Internet Protocol* (VoIP) could streamline the intelligence-sharing process and prevent future attacks.

Although such technologies are readily available and relatively inexpensive, the telecommunications infrastructure that would deliver these services to small city or rural police forces and the residences of agents who require this information is, for a number of reasons, inadequate. Broadband Internet access is available to only 8 percent of residences nationwide. The availability of broadband Internet access for small city and rural police forces is probably no better.

The Solution

Just as the interstate highway system in this country was introduced as the National Defense Highway System and was initially funded as a defense project, a comparable program could provide a majority of American households with broadband Internet access with a multitude of societal and economic benefits. Some authorities state that this universal broadband deployment, that is, broadband Internet access to a majority of American homes, could produce a consumer benefit of $500 billion annually.

The *National Defense Residential Broadband Network* (NDRBN, an acronym pronounced "nidderband") would best use wireless broadband (802.11b and associated protocols) to deliver broadband services to the "last mile," that is, the residence or small business. This would ensure redundancy and survivability in the network and would encourage investment in a wide range of technologies and businesses, creating millions of jobs.

Benefits

The NDRBN would provide many benefits for both the military and for civilians. National security applications would include:

- Allowing government security offices to interact with state and local police forces via a variety of Internet resources (web sites with current terrorism intelligence information, email alert bulletins, video conferencing, video on demand, and cost-free, long-distance voice) resulting in a streamlined process for sharing information on terrorist threats nationwide. This would empower servicemen and -women with a means of working at remote sites, including their residences, which could save the Department of Defense (read U.S. taxpayers) billions of dollars annually in telephone, travel, and relocation expenses. This would also enable military reservists to stay in touch with their units remotely via email, high-bandwidth web sites, video conferencing, long-distance calling, and video on demand.

- Allowing military veterans subject to recall to stay informed of military issues via email, web sites, video conferencing, and video on demand.

- Empowering the *Veterans Administration* (VA) to provide "e-medicine" healthcare applications for millions of veterans who currently find the VA system overwhelmed, resulting in delayed or inadequate care. E-medicine would also save the taxpayers millions annually in VA healthcare costs.

Civilian applications and benefits include

- Reducing U.S. dependency on foreign oil by empowering Americans to telecommute using video conferencing technologies. This applies equally to reducing costly business travel utilizing the same technologies. Studies have shown some 30 percent of the U.S. work force could telecommute if residential broadband Internet access was available.

- Developing and deploying e-medicine applications to improve healthcare for citizens and reduce the overall cost of providing healthcare. This network would connect all hospitals, clinics, and doctor's offices to patients statewide.

- Enhancing distance learning for both adults and school-age children via email, high-bandwidth web sites, video conferencing, and video on demand.

- Enabling an urban work force to move to less densely populated (and less costly) parts of the state. This could be an engine for rural economic development

- Boosting worker productivity by jump-starting the now beleaguered *Information Technologies* (IT) sector of the economy.

The Opportunity

Just as the National Defense Highway System is now a highly valuable part of our civilian economic infrastructure, the NDRBN could be a powerful engine for economic development.

In addition to the human capital, many vendors of telecommunications equipment are experiencing financial difficulty in this down market and offer their products at considerable savings over what they charged two years ago. Components critical to building the NDRBN could be purchased by the builders of this network at great savings relative to their costs two years ago.

A crucial element of the NDRBN is the vast network of fiber-optic cable that constitutes the backbone of the Internet. Of all the fiber-optic cable laid by telecommunications companies, no more than 10 percent (estimates vary widely, but are universally pessimistic) are "lit", that is, in use and generating revenue.[1]

This overabundance has been termed "bandwidth glut" and has resulted in the bankruptcy or near bankruptcy of many such businesses. The chief reason for economic woe in this business sector is an inability on the part of the infrastructure builders to deliver broadband Internet access to residences and small businesses; that is, supply has greatly exceeded demand in the short term. In short, a nationwide "skeleton" for the NDRBN has been built by the private sector and now lies dormant. These networks or the bandwidth they provide can also be had at great savings from financially troubled service providers.

Two potential approaches exist for building the NDRBN: The NDRBN will be government built, owned, and operated, or the NDRBN will be privately built and operated with government incentives for private sector entrepreneurs. As a government project, the NDRBN would exist as an Internet-access-only entity and provide no content or specific services of its own. In this capacity, the NDRBN would be analogous to the interstate highway system. Citizens would use the NDRBN to access the Internet in the same way they use an interstate highway to "access" their home, place of work, or business.

---

[1]"Numbers Crunched," *Telephony Magazine*, October 14, 2002: 26.

Just as the interstate highway system was built by construction contractors, the NDRBN would be built by telecommunications infrastructure contractors. Once built, an appropriate government entity would maintain the network for the common good just as state highway departments maintain the interstate highway system, albeit with monies from the U.S. Department of Transportation.

Although some might argue this constitutes government interference in the private sector, it should be asked as to why only 8 percent of American residences have broadband Internet access. This is despite the fact that most of the technologies listed earlier have been available for a number of years and that the nation is living in one of the most prosperous periods in the country's history. It can be argued that private sector entities have failed or refused to deliver broadband Internet access and it now falls to the government to provide this service for what is potentially a $500 billion annual benefit to the U.S. economy.

Another argument is that the monopolistic structure of the telephone and cable TV industries has blocked the timely rollout of broadband Internet access. These entities have had little incentive to provide this service. Cable TV companies usually have exclusive franchises to provide cable TV in their areas of service and are faced with no competition; as a result, they will not make the necessary investment in infrastructure to offer broadband Internet access. Telephone companies, despite the provisions of the Telecommunications Act of 1996, also face little competition in their specific markets and, as a result, provide less than 10 percent of their subscribers with broadband Internet access.

A compromise between public and private construction and ownership of the NDRBN would be to provide incentives to service providers to roll out broadband Internet access. This would cover the construction and maintenance of certain parts of the NDRBN. 802.11b or similar wireless technologies are relatively inexpensive to deploy and may provide opportunities for "Mom and Pop" small businesses to offer such Internet access in their neighborhoods. Most Americans first subscribed to the Internet via small, local ISPs who could get into the business for as little as $5,000 depending on the size of subscriber base they sought. Many such operators are still in business today. The same scenario could work for the rollout of the NDRBN.

The deployment of the NDRBN need not be the scene of "pork barrel" projects. Government incentives for small entrepreneurs could be limited to government-backed SBA loans aimed at NDRBN builders or tax relief for a number of years following that builder's construction of their part of the NDRBN. It should be noted here that the transcontinental railroad network would not have been built in the United States were it not for the incentive of federal land to the railroads in exchange for miles of track laid.[2]

The benefits of the NDRBN are manifold. The current environment in the telecommunications market presents a grand opportunity for our society to build this network for the common good (and common defense). New technologies make the rollout of this network both technologically and economically feasible. Like President Eisenhower's National Defense Highway System, a wide variety of options exist to fund, build, and operate this network as a partnership between the public and private sectors. The time to build the NDRBN is now.

Better Living Through Telecommunications: The Social Rewards of Wireless Broadband

Few engineering books attempt to answer the big "so what?" of a technology. What is the social impact of 802.11 on society? The following sections point to a few social ills that can be cured by better living through telecommunications.

If It Hurts to Commute, Then Don't Commute

No traffic problems exist in the industrialized world. A large percentage of the drivers on the developed world's highways are driving to

---

[2]See Stephen Ambrose's *Nothing Else Like It In The World* (Simon and Schuster, 2000) for a treatment of funding schemes for the transcontinental railroad.

offices where they work on computers and telephones the majority of their workday. Are we to believe those in the Audis and Land Rovers do not have telephones and computers at home? Why do they clog highways to go somewhere to do something they could just as well do in their homes? Why are tax dollars consumed by their demand for highways, parking garages, and other wasteful forms of transportation infrastructure that only serve to breed more congestion?

The primary direct benefits of telecommuting occur from the reduction in travel required by the employee and the reduction in infrastructure costs at the office. However, a significant secondary benefit is the reduction in congestion costs. The *Texas Transportation Institute's* (TTI's) Urban Mobility Study reports estimates of the costs of traffic congestion in 68 urban areas.[3] Its 2001 report states:

> Congestion costs can be expressed in a lot of different factors, but they are all increasing. The total congestion "bill" for the 68 areas in 1999 came to $78 billion, which was the value of 4.5 billion hours of delay and 6.8 billion gallons of excess fuel consumed. To keep congestion from growing between 1998 and 1999 would have required 1,800 new lane-miles of freeway and 2,500 new lane-miles of streets—or—6.1 million new trips taken by either carpool or transit, or perhaps satisfied by some electronic means—or—some combination of these actions. These events did not happen, and congestion increased.

Analysis of the TTI report shows that 80 percent of these $78 billion in costs occurs in only 24 cities (comprising most of the larger cities in the United States), and 90 percent occurs in 36 cities. In Los Angeles, traffic congestion imposes estimated costs of $1,000 per person per year.

Various studies estimate that 20 to 40 percent of jobs permit telecommuting at least part of the time. If we assume that 30 percent of jobs permit telecommuting an average of 20 percent of the

---

[3]The hourly value of $12.40 is used in the congestion analysis by the Texas Transportation Institute in its urban traffic studies discussed here. For more information, see the Texas Transportation Institute's "The 2001 Urban Mobility Report" (The Texas A&M University System, May 2001, http://mobility.tamu.edu/).

time (1 day per week or 50 days per year) and that the average commuter trip for a telecommuter is 20 minutes, then we can calculate the potential savings in travel costs and congestion. The savings in travel time are

$$\text{(180 million civilian labor force)} \times \text{(30\% possible telecommuters)} \times$$
$$\text{(33 hours/year)} \times \text{(\$6.20/hour)} = \text{\$11.1 billion per year}$$

Similarly, the savings in travel costs are

$$\text{(180 million civilian labor force)} \times \text{(30\% possible telecommuters)} \times$$
$$\text{(450 miles/year)} \times \text{(\$0.3/mile)} = \text{\$7.3 billion per year}$$

Summing up, the potential savings from telecommuting are $11.1 billion per year travel time for telecommuters, $7.3 billion per year for travel costs, and $4.7 billion per year for reduced (external) congestion costs, yielding a total savings of over $23 billion per year. Note that this is not an estimate of the savings from the accelerated deployment of broadband access; this is an estimate of the total transportation system savings from the widespread adoption of telecommuting. Assuming that these savings grow at a rate similar to the general growth rate we assume for the economy, these savings could be as much as $30 billion in 10 years.[4]

Why do commuters persist in following an Industrial Age habit in the Information Age? Sadly, we have translated the drill press to the computer. Instead of a factory, we now have paperwork factories in the form of downtown office buildings and outlying office parks. If our white-collar workforce could only break out of their Industrial Age lockstep mentality and work at their homes or local teleworking centers, a meaningful percentage of our traffic would simply evaporate.

One objection to this solution is that white-collar workers need face-to-face contact to coordinate their work. True, but does it have to be exactly at 8 A.M. through 5 P.M. Monday through Friday? Wouldn't attendance at biweekly staff meetings suffice, the rest of the time being spent at home offices or teleworking centers closer to home?

---

[4]Robert Crandall and Charles Jackson, "The $500 Billion Opportunity: The Potential Economic Benefit of Widespread Diffusion of Broadband Internet Access," Washington, DC, July 2001, p. 37.

Another objection is the need for socialization. First, one would speculate that if one's only friends are those in the office, that individual should work at "getting a life." Secondly, does that socialization have to be all day Monday through Friday? We can take our cue from America's first home workers: farmers. Much of rural day-to-day socialization evolves around a visit to a neighbor for coffee or a stop in the local café. This practice could be adapted in suburbia.

A more concrete objection, especially among high-tech workers, is the need for expensive computers with high-speed connections to the Internet. Many corporations who have weighed the cost per square foot of maintaining workspace in downtown office spaces or out at the local tech center found it cheaper to send their people home with that expensive computer and offer the employee a stipend for his or her home office. Competition to offer high-speed Internet access to the home by phone line, cable TV, or wireless would negate this Internet bandwidth objection.

So is there a high-tech, expensive solution for our so-called traffic problem that elected leaders can plan and impose over the next few decades? The best thing elected leaders can do is *nothing*. Once traffic becomes enough of a painful experience, those that don't *have* to brave traffic morning and night simply will not do so. They will stay home to do their work. They will slowly realize that the Information Age is driven by ideas. The formation of ideas is not limited to the hours 9 to 5 Monday through Friday. Nor are ideas limited to one place geographically, especially not a corporate cubicle. Perhaps we will ask that question from World War II gas-rationing pitches: "Is this trip really necessary?"

Concrete roads were first developed by the Romans to move their chief commodity, labor (soldiers and slaves), quickly and efficiently throughout the empire. In the Information Age, our chief commodity is ideas. Ideas don't need roads for their transportation. Ideas move well over fiber-optic cable, phone lines, coaxial cable TV cables, or even the airwaves. We must ask our leaders to compare the cost per mile of pavement (millions of dollars) to fiber-optic cable (approximately $25,000) or even wireless broadband. Its time to tell elected officials that they need to plan for the 2020s (the Information Super-highway), not the 1920s (the Lincoln Highway).

Affordable Housing Is Where You Find It

A societal ill dictates that good-paying jobs are only found in downtown high rises or in outlying office parks. As that is "where the money is," white-collar or Information Age workers strive to live within an easy commute of those offices. This in turn breeds congestion as everyone attempts to spend exactly the hours of 9 to 5 Monday through Friday in those offices.

Most government officials have not caught on to this yet, but new telecommunications technologies have largely negated the need for white-collar or Information Age workers to drive to an office. In the bad old days, there were factories. The machinery in a factory cost a lot of money. Workers had to go to the factories because that was where the jobs were. With the coming of computers and the Internet, white-collar and Information Age workers can work from their homes in the suburbs. As telecommunications improve in more rural areas, white-collar and Information Age workers can do their jobs from much farther afield and improve their net worth.

The average sale price of a single-family home in Denver, Colorado, at the time of this writing (spring 2003) is over $275,000. Sale prices of single-family homes in small-town Iowa, by contrast, hover at $30,000. Imagine a white-collar or Information Age worker making his or her big-city income in a low-cost area like small-town Iowa. The savings in mortgage payments are obvious. If one went over his or her monthly budget, other significant savings could be found in such a lifestyle change (telecommuters don't rack up as many miles on their cars). By saving as much as possible of that big-city income while living in a low-cost area, financial independence could be realized in a short period of time. Contrast that with the "I owe, I owe, it's off to work I go" consumptive/dependent lifestyle of the suburbanite.

Imagine a society where we didn't have to move every few years to stay competitive in our work. Imagine being able to stay in one place most of our working lives as our work was not dependent on relocating to another city to take another job with the subsequent loss of contact with friends and family. Information Age work is not dependent on working from a specific cubicle in a specific office building. It is dependent on having a consistent marketable skill. In theory, an Information Age worker should be able to change jobs or contracts repeatedly over a working life and never have to change residences. Imagine successive generations living under one roof. No mortgage, no moving expenses, no closing fees.

Family Values

Why are there latchkey children? The answer is simple: because some people think that the only place a job can be done is in a specific office at a specific time. Therefore, Mom and Dad must be at those specific places at those specific times, which leaves their children coming home from school to an empty house and, perhaps, trouble.

If it were better understood that most white-collar jobs could be performed from a home office, then moms and dads would not need to be absent when their children arrived home from school. Hence, social ills associated with absent parents would be greatly diminished.

The Role of 802.11b in Better Living Through Telecommunications

At about 1920, half the American population either lived on farms or in farming communities. Today that figure is less than 5 percent. What happened over the last 80 years? Obviously, the farming half of the population moved to the cities in search of better pay. Cities have always been communication hubs (railroad, highways, telephones, and the Internet). Also, it has historically been perceived that cities contain more vibrant communities (theater, opera, live music, and cinema) than what could be found in the country.

The telecommunications revolution can make both of those perceptions obsolete. Historically, rural areas are the last to get the newest telecommunications technologies, as they pose a lower rate of return for the service provider. Much of rural America's telephone and electric infrastructure was built by cooperatives of farmers pooling their labor and money to install telephone and electric poles and wires. This scarcity of telecommunications infrastructure extends to the suburbs of the largest cities in the United States. As of 2002, only an estimated 8 percent of U.S. households had access to broadband services. This is due largely to a lack of infrastructure for broadband and perhaps a reluctance on the part of incumbent service providers to make the investment on DSL infrastructure when they could be retaining the cash to make their stock more attractive or pay down mountains of debt.

802.11b and associated wireless protocols, by virtue of being simpler, smaller, cheaper, and more convenient for competitive service providers to deploy, enable more service providers (ISPs, *application service provider* [ASPs], power companies, municipalities, cable TV companies, and wireless service providers) to reach more customers. Most importantly, this would allow data service providers to offer voice and video services. Voice services offer higher margins than data services. The prospect of generating revenue from both data and voice should motivate service providers to enter as many markets as possible. This trend could be enhanced where service providers can bypass the facilities of incumbent service providers. Even if alternative service providers don't specifically charge for voice, bundling it with data could prove attractive. The same is true of offering wireless video over IP as a means of bypassing cable TV company monopolies.

Wireless broadband is a technology that can ignite competition for telecommunications services. What little competition the Telecom Act of 1996 inspired in the telecommunications field was for business services in business districts throughout the United States in the late 1990s. Competitive service providers sought to "cherry pick," that is, offer service only to business subscribers, in high-density areas, thus maximizing *return on investment* (ROI) in the infrastructure in city centers. This ultimately led to hypercompetition, a "race to the bottom" in pricing, and ultimately the shakeout and bankruptcy of many service providers.

This hypercompetition should have the effect of driving service providers to find less competitive markets (suburbs and rural areas) where they would enjoy less volatility. In some metropolitan markets, the price of a data T1 has dropped to $300 from precompetition levels of $1,200 per month. The deployment of 802.11 with 11 Mbps at a $100 per month subscription rate would torpedo PSTN data

Figure 10-1
The transition
from a PSTN to
an Internet
infrastructure

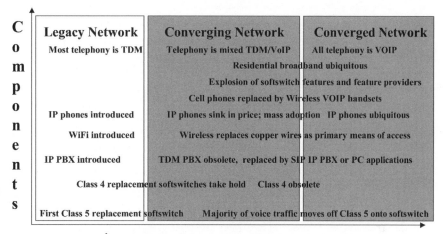

Softswitch Replaces PSTN Timeline

Legacy Network	Converging Network	Converged Network
Most telephony is TDM	Telephony is mixed TDM/VoIP	All telephony is VOIP
	Residential broadband ubiquitous	
		Explosion of softswitch features and feature providers
	Cell phones replaced by Wireless VOIP handsets	
IP phones introduced	IP phones sink in price; mass adoption	IP phones ubiquitous
WiFi introduced	Wireless replaces copper wires as primary means of access	
IP PBX introduced	TDM PBX obsolete, replaced by SIP IP PBX or PC applications	
	Class 4 replacement softswitches take hold Class 4 obsolete	
First Class 5 replacement softswitch	Majority of voice traffic moves off Class 5 onto softswitch	

2002   2004 2006 2008 2010 2012 2014 2016 2018 2020
Time

pricing and possibly much of the PSTN market share in both voice
and data services (see Figure 10-1).

Conclusion

The spread of broadband Internet access to a majority of U.S. households will probably happen in the form of 802.11. The demand for
broadband will have the effect of bringing different forms of delivery
(DSL, cable modem, and 802.11) into the marketplace. The form of
access that is least expensive and most easily deployed will win.
That form of access is 802.11.

INDEX

*Note: Boldface **numbers** indicate illustrations.*

ABOUT THE AUTHORS

Frank Ohrtman has 20 years experience in professional networking, including high-security installations for industry and the military (US Navy). A former Account Manager at Lucent Technologies, Netrix and Vsys, he is most recently the author of *Softswitch: Architecture for VoIP*. He has a master's degree in Telecommunications Engineering from Colorado University. He lives in Denver, Colorado.

Konrad Roeder has over 20 years experience with various technologies such as Wi-Fi, GSM, Cellular, SMR, VoIP and ISDN. Konrad has a BS in Computer Science from University of New Mexico and MS in Computer Engineering from Florida Atlantic University. He has 3 patents issued and 3 patents pending. He currently works at T-Mobile on the 802.11 project deploying Wi-Fi in Airports and Airline Clubs. Mr. Roeder lives in Seattle, WA.